Technical Report Writing

and

Style Guide

Technical Report Writing and Style Guide

How to write even better technical reports

Tony Atherton

ATC

Library of Congress Control Number: 2018675309

ISBN: 9798674111979

To my wife, Judy.
For putting up with the too many hours I spent writing this book.

ACKNOWLEDGEMENTS

Like many of us, it is impossible for me to single out individuals who over the years have influenced my learning about the skills of writing and report writing in particular. Many work colleagues have helped me to learn from errors in my writing. Many of the over 5000 delegates who have been on my various writing courses have sent me scurrying to reference books with their unexpected questions. But, above all, I am indebted to the authors and editors of the many reference books I have learned from and am still learning from, sometimes with a bit of a struggle. Many of them are listed in the Bibliography at the end of this book.

CONTENTS

PREFACE

I have trained engineers in technical report writing, mainly in the UK, for about 20 years and had long thought that one day I would write a book based on the course material. The coronavirus pandemic of 2020 provided the opportunity.

The course has been well received in both small and large organisations (I know, what else would I say?) and there have been loads of repeat orders.

Here are a few comments – no one wants to read lots of them. Two are from delegates and three are from managers who arranged for me to train their staff.

Delegates:

'I keep the workbook on my desk for easy referral.' – Southern Water

'I regularly struggle with some of the ideas he talked about so it was a great help.' – Pilkingtons

Managers about their delegates:

'His reports immediately took on a completely different look and feel.' – Synthomer

'Two of the engineers even went back and looked at their reports with a view to rewriting them.' – Transport for London (TfL)

'His report has sailed through editing with hardly any comments at all, which he is very pleased about!' – The Health and Safety Laboratory (part of HSE)

I hope the book will be as useful to you as the course has been to so many engineers.

1

INTRODUCTION

This book is based on a training course about writing technical reports that I have presented over 200 times to more than 2000 engineers over a period of about 20 years. The course itself was based on my own experience of engineering (I am a retired Chartered Engineer), my experience of writing reports (both technical and business) and my experience as an author of nearly 100 articles and four books.

Over the years, many managers have told me much the same thing about why they wanted the training for their engineers. Here are a few of their comments about the problems they saw with reports in their organisations.

> 'The main issue is that the reports are not structured in an efficient and coherent way, all the information is there but the reports tend to be all over the place... The main aim for my people is to understand how to build a case based on the gathered information and evidence, and write that in a fashion that is clear, precise and structured without any added "fluff".'

> 'The reports are lacking in detail, poorly constructed and sometimes contain contradictory concepts within one document.'

> 'We need to be able to show that we have considered every angle of a problem, documented our thought processes in a crystal-clear manner and have reached the true root cause (not a cause which itself has a further root cause – i.e. lazy thinking).'

> 'Here are some of the problems: understanding who the recipients are and the level the report needs to be pitched at, capturing key aspects in the Executive Summary, getting the concept of the report into the introduction and understanding the scope of the report.'

Others have complained about ambiguous, misleading or rambling statements; colloquial language; inconsistent terminology; not checking that all abbreviations are listed, or that those that are listed are used; not reading the report before presenting it for review and not checking spelling.

Oh dear! The main problems are common and we seek to solve them in this book.

I often compare a report to a machine. Machines are designed, built and tested for a defined purpose – they must do something and do it effectively and efficiently. So too with technical reports. They too must do something and do it effectively and efficiently. Often, they have to convince someone to take some recommended action. As with machines, they should be designed, built and tested, or in writing terms – planned, written and edited (and reviewed).

So, what is a report? A technical report is a clear and concise account of an investigation, or part of an investigation, and its outcome. That account is very important to some people and may guide decisions they have to make. They have asked a question and they want an answer. Notice the phrase, 'clear and concise' – two vital qualities of good reports.

Reports are also your products, especially if you work for an engineering consultancy. You may not work for a manufacturer, but consultancies manufacture reports – and your reports have a direct impact on your reputation and that of your organisation. They can create a lasting impression on clients, suppliers, the public, the media and on anyone else who reads them, and that can include your manager, senior managers and board members. If reports are good, they can help to build client loyalty. If they are bad, they can damage it.

What makes a good report? The most obvious requirement of a good report, which we should be able to take as given, is that it answers clearly and unambiguously the question that was asked. We can hope that at least the writer thinks this is true of his or her report, but what about the readers?

From the readers' perspective they need more than an answer, they need an answer they feel they can trust. So, there are many more things, not just an answer, that a good report has to provide and amongst them are three big ones.

The first of these is that the readers can see quickly that it is technically robust, which helps them to start to trust it. The second is that the readers can follow the reasoning easily, which comes from the way you arranged or sequenced the information. That also builds trust. The third is that the intended readers can understand it easily. That is, the quality of the writing is good enough to keep them reading and good enough not to distract them from the message. They understand all the words – and the paragraphs and sentences do not meander on and on. Your readers are busy people and will resent poor writing.

All three principles focus on your readers' experience, not yours. If you write something this way or that way – how will it affect your readers? Keep asking yourself how this or that will affect your readers. Will they understand this bit? Is that the best way to describe that bit? Try to have specific readers in mind as you write and edit, but you may need to think of different people for different parts of the report because many people do not read an entire report. For instance, senior managers may not read the heavy technical parts of your report. We will look at how to write for readers with different levels of technical knowledge later.

If you do not know your readers, try to imagine what they are like. Aim your writing at them as best as you can. Are they all technical experts? Probably not. Are they all decision makers? Again, probably not. But if the decision makers are not technical experts, they will ask those who are for their opinions of your report. Those opinions will influence the decision makers. So you should always write for your readers, not for yourself – and think about what that statement means. It is the readers who are the most important people here. It is great if you feel good about your report, but it is even better if the readers feel good about it.

The first of the three principles, the robustness or technical accuracy of the content, lies outside the scope of this book. I trust your judgement on that score and have taken it as given. Being technically robust is one thing but convincing your readers of that is something else. Convincing them starts with the aim of the report. It builds as you describe your procedures, but more of that later.

All reports have an aim and, as many technical reports seek to answer one or more questions, phrasing the aim as a question or set of questions can be helpful. It makes the relevance of the report immediately obvious to the intended readers. What caused this accident and how can we minimise the chances of it happening again? What is the best design for this new road intersection? What will be the flood risk if we build on this site and how can we mitigate it? How will various types of packaging affect the shelf life of this new medication? There is a problem; how do we solve it? And so on: there are endless questions and answers, and endless reports.

The second principle is to sequence your information in a logical way that the readers can relate to, so that they can follow it instinctively. That starts by using the standard sections that your readers expect to see, which are the executive summary, introduction, procedure and so on. It is likely that you will have little control over these as your organisation's templates may dictate them to you. A technical report is not the place for experimental arrangements although some technical reports do use other arrangements, such as the managerial format or the 1:3:25 format – both of which we will look at later.

However, sequencing does not stop with the format. The parts of your report that you do control also need a logical sequence. These include the subsections that you introduce, the paragraphs within them and even the sentences within the paragraphs. We will look at them all.

The third principle is to write well, or at least well enough not to trouble your readers. Many people have said that good writing is hard work, and it is. Good reports do not just happen. An effective approach is to plan the report overall and then plan more detail section by section, then write it section by section and finally edit it section by section. Plan, write, edit! It is the equivalent of designing a machine, then building and testing each part before bringing it all together. Good planning, or good designing, reduces the amount of time you are likely to spend rewriting, but there will always be some rewriting and sometimes quite a lot. This is not a production line so you may never get it 'right first time', but this book will help you to get it better first time – and that will save you time overall.

Consistency of grammar and style is another important principle to follow. Some of your readers will notice inconsistencies and anything they see as negative is bad news. Your organisation's template will dictate consistency for major things, such as the page layout, fonts and heading styles. But what about the relatively minor things like how to punctuate bullet points, how many spaces to leave between sentences (one space is preferred), and when to hyphenate words and when not to? For those things you need a style guide (and a dictionary). While many organisations provide templates, relatively few provide a style guide. That is why this book includes an style guide with more than 130 style issues for technical report writers.

Ideally, every report from your organisation should have a consistent layout and style – just as they all use the same logo and organisation name. It is part of your organisation's brand or image. Without consistency, different authors make different choices and consistency is lost. This is particularly important for engineering consultancies that produce hundreds of reports a year from dozens of engineers.

Do readers notice inconsistencies? Many do and noticing may become a slippery slope down which a few slide into asking, 'If you didn't take the trouble to get those things right, how much trouble did you take to get the difficult technical issues right?' At that point, those readers may start to lose trust in you and, while their views may be pedantic, they may influence others. Even though that is a small risk, we want to avoid it especially with external clients.

As we all know, there are some differences between British and American English. We are fortunate that the rigours of writing technical reports impose restrictions, especially on terminology, that make it unlikely that the readers from one tradition will struggle to understand a technical report from the other tradition. As the managers quoted earlier found, it is things like a poor structure, lack of clarity, muddled thinking, not recognising what matters to the readers and so on that are far more likely to hinder understanding than the small differences between British and American English. Nevertheless, there are some differences and they could be important for some writers, so they are mentioned in the text where appropriate, especially in Chapter 8, the Style Guide.

As mentioned earlier, this book is based on a long-running technical report writing course and it embraces the three principles mentioned here (technically robust, a logical structure, and good English), but it has a lot more besides. I wrote it to help you to write shorter and better reports and to write them faster, and to some extent to make the task easier. Writing a good technical report is never likely to be easy, but it does not have to be as difficult as it sometimes is. The book will help you to make good choices where there is more than one way of doing things, and then apply those choices consistently. Obviously, it must never override any templates or style guides your organisation tells you to use, but it will help you to be consistent in many areas where you have no other guidance.

Each chapter looks in detail at the various aspects of producing a technical report. There are chapters on planning or designing reports, on writing them and on editing them. The chapter on editing includes a look at using the Microsoft Word grammar checker and its built-in readability statistics (which many do not realise are there). There is also advice on proofreading. Another chapter examines in detail the standard sections normally used in technical reports, still another is a reference chapter about our scientific units and the rules to follow when writing them. Finally, the style guide provides guidance on over 130 style issues.

Most engineers did not choose a career in engineering because they ached to become authors, but once you start writing reports that is what you have become. Perhaps writing a technical report will never be a lot of fun, but it does not need to be a lot of misery.

To summarise this opening chapter, your report should:

- Have an aim and achieve it, maybe as specific questions and answers.
- Be technically robust.
- Be sequenced so that your readers can follow the reasoning easily.
- Be written in English that is good enough not to distract your readers from your message.
- Have a consistent style so as not to distract your readers from your message.

2

PLANNING

At its simplest, a report is a message from an author to one or more readers. Put like that it is obvious, but there are three parts to that statement and an effective report is one where these three parts work well together: the author, the message and the readers.

The importance of this trio has been recognised since the time of the Ancient Greeks and is studied in rhetoric. In fact, the trio are known as the three rhetorical appeals – three things that need to work if a message is to be persuasive. Some engineers may not give it much thought, but a report does have to be persuasive if the readers are to carry out the recommendations. The three rhetorical appeals are called ethos, logos and pathos – not that you really need to remember their names.

Also in this chapter we will look at:

- Why you should design your report rather than just write it, the benefits of doing that and how to do it.
- The importance of agreeing the report's aims as early as possible.
- How what you know about your readers might influence the design and the writing.
- The information you want to pass on and how to do it, facts versus opinions, and the inter-relationship between the text and illustrations.
- The standard format for technical reports and some alternative formats.
- How to sequence the information in three layers – the sections, subsections and paragraphs.
- How to use bullet points to plan the sequence.
- Choosing the heading styles.

2.1 – Three Rhetorical Appeals

The first of the three rhetorical appeals, *ethos*, is about the author. Why should I trust you? What is your authority for writing this? Why should I listen to you? For most technical reports this rhetorical appeal is relatively easy to achieve – provided you are aware of it.

The chances are you have been asked to do this work and write this report, or your organisation has. Maybe it is part of a contract that your client has awarded to your organisation or maybe it is an internal request. That all adds to your authority and any report written by a consultancy for a client should mention the contract in some way, but there is more to it than that. When you describe the procedures or methods you used to gather your data or results, you are using the ethos rhetorical appeal. 'Look!' you say, 'See how I did it. You can trust me, I did it right. I know what I am doing.'

The technical accuracy of the report (or robustness as we put it in Chapter 1) is also part of the ethos. Consider the opposite, when readers question how thorough the work was, when they pick holes in the methods you used – then you lose credibility. Also, be honest. If you have contrary evidence – describe it and explain why you have discounted it (if you have). Look after the ethos part of the three rhetorical appeals, plan for it – do not leave it to chance.

The second of the rhetorical appeals, *logos*, is an appeal to reason – to the logic of your argument. Does it make sense? Is it logical? Does it hang together? It is again about robustness but this time it is about how robust your reasoning, argument or case is. Again, it is more than that. Yes, your case must be logical, but also it must be seen to be logical. Here we are focussing on the arrangement or sequence of your argument. This goes from the arrangement of the major sections to the subsections, to the paragraphs, to the sentences and even to the words you choose. Once again, do not leave it to chance – it needs to be planned. Strangely, considering they are written by engineers, the logical arrangement of the details in technical reports is one of the most heavily criticised aspects of such reports.

The third appeal is to *pathos* – the appeal to the readers, specifically to their emotions. Do I care? Does it matter to me? Why should I be interested in your report? Answering that starts when you spell out the aims of the report, which maybe they gave to you anyway. They are interested because they asked you to do this; they have a problem and they asked you to solve it for them. Do not be reticent about gently and very carefully reminding of that through how you express each aim: this aim is to solve this problem so that you get this benefit. You are reminding them of what is in it for them. Of course, they are looking for solutions to their problems – but what they really want are the benefits they will get from those solutions. Once again, do not leave that to chance, plan the pathos.

2.2 – Design your Report

Reports consist of sections, which consist of subsections, which consist of paragraphs, which consist of sentences, which consist of words – and in really good reports every one of those is the right choice in the right place. That happens by design, not by chance.

Imagine you wanted a house built. Would you invite a builder along and say get started; build me a house right here? No! If you did, a few weeks later you would be telling the builder to move that wall from there to here, make that room bigger, move the kitchen to there – a disaster. Instead, you would hire an architect and make rooms bigger or smaller at the design stage. Only when the design was complete would the builder become involved.

Yet, too many report writers seem to skip the design stage and then waste time moving their metaphorical walls around.

There are three stages to producing a good report. The first stage (assuming you already know what the aims are and who the readers will be) is to plan the report in considerable detail until you have a workable design. No doubt some of that design will change later, but not by much. The second stage is to produce the draft by thoughtful writing whilst following the plan – by now you have a rough diamond. The third stage is to edit the draft carefully until it is fit for purpose – by which time you have a sparkling gem.

Without a good design based on thorough planning, you will waste a lot of time writing and rewriting. However, even with good planning, you will still need to do some rewriting because not everything will be right first time. I would advise almost everyone to increase the amount of time they invest in planning a report because they will save much more than that when writing and editing; at least after some practice.

A good report addresses your readers' issues and concerns (pathos), which means you have to know what those issues and concerns are when you are planning your design. The report should tell your readers what they need and want to know, which may not be the same things, and it should answer their questions in a way they can understand, believe and trust. It should also tell them about other things that they may not have thought about; things that you realise they need to know to get to sound conclusions and recommendations. That is true even if some of these facts may not be good news. That is part of the honesty that must run through a technical report, based on your integrity as a technical expert – part of the ethos.

A good report does all of that clearly and succinctly, and with conviction, because you designed it that way. The intended readers understand it easily (pathos) because you wrote it clearly using language and terminology they know, and the information is sequenced in an obvious and sensible way (logos). Each point follows the previous one in order and leads logically to the next. By the time they read your conclusions, those readers who read the whole report are ahead of you in knowing what they are. There are no surprises in the conclusions – because you designed the report that way.

Can you write like that so that every one of your intended readers can follow and understand every bit of your message the first time they read it? No? Me neither, but with good planning we can get close if we try, although we may never fully achieve it. Many have said that failing to plan is planning to fail and there is a lot of truth in that adage.

There is a lot of truth in another adage too, that easy writing means hard reading. The reverse is also true. Hard work by the author makes the reading easier for the reader, but that takes time. The chances are remote that you or I can write a perfectly organised message over many pages in succinct yet flowing English without a lot of effort. Writing a good technical report takes time: time to plan, time to write, time to improve, time to edit, time to review and time to proofread. All this comes after you have collected, organised and analysed all your data, and drawn and tested all your conclusions and all your recommendations.

Once you have improved a report sufficiently to make it good enough to achieve its aims it will be time to move on to your next piece of work. You could always improve your report a bit more here and there but there is a law of diminishing returns. Move on.

Good planning is the easiest and most reliable way to reduce the time taken to write well, to get all the right sections, subsections, paragraphs, sentences and words in the right place. Never get the builder in before the architect has designed the plan; never start writing before you have designed your plan.

2.3 – Aims

Defining the aims correctly is critical and defining them before you start work is very important. Getting them even slightly wrong risks the response that, 'The report is not quite what we wanted.' The worst-case scenario is that the report is rejected and that could mean losing the client to a rival if you work for an engineering consultancy. At the very least, getting the aims slightly wrong or late risks a lot of reworking and rewriting. Many report writers are given the aims, which are likely to be identical (or almost identical) to the aims of the investigation itself.

As already mentioned, consider stating the aims as a question or several questions. We can call these the 'What' question or questions. Many reports do this and it can be very effective. Questions are asked and the report answers them.

Consider this example from a report commissioned by the UK's Chief Scientific Adviser (the report was published in 2012).

> *'What are the major risks associated with hydraulic fracturing as a means to extract shale gas in the UK, including geological risks, such as seismicity, and environmental risks, such as groundwater contamination? Can these risks be effectively managed? If so, how?'*

The two-page Summary of the resulting report answered the three questions very briefly in its 73-word opening paragraph before providing more details in the rest of the Summary. It was a well-designed piece of writing, whether or not you agree with the report's full answers.

Even with internal reports, seriously consider agreeing the aims formally, even signing and dating them. That will deter you from wandering off target and will defend you against anyone who moves the goal posts later. Many internal reports are prompted by verbal instructions, which can easily be misinterpreted. Email what you think are the intended aims to whoever has asked you to do the work and, with a little toing and froing, you will reach agreement – in writing.

The aims may be split into sub-aims, which are usually called objectives. People use several terms quite loosely for the aims: purpose, objectives, terms of reference and scope. For the moment we will use 'aims' but we will discuss these different terms fully in the next chapter.

When writing the aims, ask yourself three things:

- What do I want my readers to know that they did not know before?
- What do I want my readers to do about it?
- How do I want my readers to feel about it?

The third question may seem odd, but readers are affected by what they read (pathos). The least you want is that they respect the report, see it as a professional job and are satisfied. However, do you want more than that? Do you want to instil a feeling of urgency, for example, or reassurance? You may need to think carefully about how to achieve such feelings and plan how to do it.

The aims are normally part of the introduction and appear after the background and before the methods. In a few long and complex reports, they may become a separate section after the introduction.

The background describes the problem and its significance, giving the reason or motivation for doing this work and your mandate for doing it. It answers the 'Why' question whereas the aims answer the 'What' question.

Answering the 'Why' question may be imperative if your report enters the public domain, especially if it was paid for from public funds.

The methods describe how you obtained your results – the 'How' question, possibly the 'When' question too. Other technical specialists must be able to recognise your methods as valid and reasonable. One hidden objective here for you is to establish your credibility (ethos) – you did it the right way and are above criticism. In long reports, the methods may break away from the Introduction to become a separate section.

2.4 – The Readers

Without readers, there is no point in writing. Your job is to write a report that meets the needs of your intended readers, as well as your own. They should be able to understand it easily, without a struggle. That cannot happen unless you know something about them or can make sensible assumptions.

Think about who your likely readers are, the people you are writing for. What will they want from the report? What do they need from it? The two might not be quite the same. Also, there could be things they ought to know about but are unaware of, but you are – things you should tell them. After all, you are the expert here. If you were buying a house, you would expect your surveyor to tell you about problems that you never thought about; so too with reports.

What do they already know about this investigation and its background? Should you remind them of some or all of that background, or none of it? How will you judge that? Even if most of your readers are aware of the background you should remind them of the most important points. By so doing, you can justify the time and effort that have gone into this work – you are answering the 'Why' question. In the past, parts of the press have criticised some health and safety reports, seeing them as a waste of time and money. If your reports will be made public, justifying your work can help to avoid unwarranted criticism.

How technically competent are your readers in this subject and do they understand the normal jargon? Almost certainly you will need to use technical jargon in the main body of the report and some of your readers will be familiar with it, but not all. Equally, you should avoid as much jargon as possible, even banning it, in the parts read by non-technical people – the executive summary, conclusions and recommendations.

Now a strange question: is it possible that some people will pull your report out of some archive in 10, 20 or even 30 or more years' time? If this is possible, then consider including an appendix with a lot more background than people of today will need. In the future, many people will know almost none of the background. Sometimes, people need to read reports that were written before they were born.

Throughout your work, whether planning, writing or editing, be aware of your readers. There will be many times when you have to decide how to say something or how to sequence some information and wonder which of several ways is best. The best way is the one that is best for your readers and sometimes that may have to be your best guess. As this is a report (not fiction – I hope) use another adage: write for your readers, not for yourself.

Few people read it all

When you buy a magazine, do you read it all? I guess that most people do not read every article in a magazine and it is likely that many people will not read every part of your report. Do not be disappointed by that statement because it is good news. It enables you to reach both technical and non-technical people with the same report. Senior managers, unless they are particularly interested, may only read the executive summary, the conclusions and recommendations. Technical specialists are most likely to read all of it, partly because they are genuinely interested and partly because it is their job to do so. Then they will tell senior managers whether they think it is reliable or not. Even technical specialists may not read all the appendices, but they might dig into parts of them for reference.

This simple fact, that many people read parts but not all, gives us a neat solution to the usual question of how to reach both technically qualified and non-technically qualified people with the same report. The answer is to write for the readers but with the reasonable expectation that you can mainly aim at different readers in different parts of the report.

Almost everyone will read, or at least scan, the executive summary, conclusions and recommendations so write these sections, the most widely read, in as plain English as possible with as little jargon as you can manage, or even no jargon at all. For example, some organisations ban all acronyms in the executive summary, even obvious ones like UK, USA and EU.

Many people will read the introduction so avoid jargon there too if you can, especially when stating the aims and describing the background. The methods or procedure section will probably have fewer non-technical readers than the background, so more jargon is permissible.

The going gets heavy in the results and analysis sections and the jargon can mount up. The people most likely to read these will be fellow specialists and for them the jargon is everyday English, even if most other people would wonder what you are talking about.

If you have an analysis section the very title will scare away most non-specialists but tread carefully if you have a discussion section. Now the title is warm and cuddly, and more people will be tempted to read it. Some of these readers will not understand much of the jargon, if any, and you must stick to words they do understand.

As already mentioned, the conclusions and recommendations will have many readers and you are, once again, writing for a wide audience using as plain English as you can.

If you use appendices, then these are most likely to be read, or even just looked at, by technical specialists. Few non-specialists will venture here except maybe from curiosity. Once again, jargon will come to the fore.

Keep these thoughts in mind as you plan, write and edit. Your readers are a mixed bunch – and by adjusting some of your writing style (we shall see how later) you should be able to keep most of them happy for most of the time.

2.5 – The Information

The information is the main part of the logos element of the three rhetorical appeals – the logical arrangement of your argument. Once you have settled the aims and have thought carefully about your readers and what they will be able to understand, the next step – and it is a very big step – is to organise the sequence in which you will present your information to your readers.

One of the most frequent moans about technical reports is that the intended readers cannot understand them. This is not just about the jargon, although that is part of the problem, it is also about the fact that they struggle to follow the reasoning. We hinted at this earlier, they cannot see a logical pattern to the case (assuming there is one). That usually arises because the information is poorly organised, something you should fix at the planning stage. As one of the managers I mentioned in Chapter 1 put it, 'all the information is there but the reports tend to be all over the place.'

A lot of the time you spend on planning should be about organising the sequence of your information. Getting the sequence right for your readers is vital and it is a lot easier to do when you are not simultaneously writing the text.

There are at least three layers to this sequencing. The highest level is to sequence your major sections and this will probably follow standard sections given by your organisation's template. As you progress you will sequence the second layer, the subsections, and then the third layer – the information within the subsections. The second and third layers will not be in your templates, you will sequence them yourself.

Start and finish with a bang

People often suggest that you should start and finish with a bang. While that is true for novels it has an element of truth for reports. In novels, the aim is to hook the readers at the start to encourage them to read on. That is also true in reports but with the addition of convincing them that this report is actually important to them.

The most important points in a traditional report are at the beginning and the end – the question and the answer – so starting and ending well is important. For many readers, the middle with all its specialist details matters less although it is vital to others. In other words, the parts that matter most to most of your readers – and are likely to have the greatest impact on them – are the executive summary, conclusions and recommendations.

This 'start with a bang' idea of the beginning being particularly important is a guideline that is worth remembering and to some degree it applies to individual sections, subsections, paragraphs and even to sentences. However, remember that it is a guideline not a rule.

First things first

A generic guideline for sections, subsections, paragraphs and sentences is to put things in order of importance: the most important first. A tip from journalists can help. Read some reports in the better newspapers and notice how you can stop reading at any point in the report but still have a coherent story; you will know the most important points. Use that as a guide – arrange things in order of importance, most of the time.

Facts v opinions

A technical report must be honest. You will have facts, information about the facts and your professional opinions about what they all mean. Keep facts and opinions separate. This is important. Your readers must have no doubt in their minds about whether they are reading facts that you have gathered (and that includes factual information about your facts, such as how measurements were obtained and their accuracy) or your professional opinions about what those facts mean. Once they are confident that you have done a professional job, they will then want to know what your professional opinions are. For many, your professional opinions are the most important part of the report. That is what is most important to them – your conclusions and recommendations, plus the discussion where you may have explored problems or different options.

Some people misguidedly claim that there should be no opinions in a technical report, but there are. Sometimes there are a lot of them. Your readers want your professional opinions as to what it all means.

Illustrations, tables or text

How will you convey to your readers the information you want them to know? Text is certainly not the only way to present your findings to your readers and at times it is not the best. In technical reports you are likely to use both illustrations and tables. The illustrations can include diagrams, graphs, photographs and so on, and they all scream importance because they all stand out from the text. Stretching the theme a little, bullet points also stand out from the text and grab the readers' attention so, in that sense, they are a bit like illustrations.

Text, illustrations and tables should present a mutually supportive message to your readers and you should consider them together at the planning stage. Often, one is the main source of some specific information while the others play supporting roles. Which does which can change many times in a report, but you always need to know which is taking the lead. For instance, a graph can convey meaning to an informed reader much faster than text and here the text will support by explaining the main points to the reader. A graph may illustrate trends (hence illustration) better than text and much better than a table, but a table can give detailed data which a graph or text would struggle to do. Decide which is taking the lead and which is supporting. Sometimes, a graph can put the principal message across while a table provides detailed data in an appendix.

You must refer to every illustration and table in the text. No illustration or table should be allowed to float around without some text describing it. Ideally, place the illustrations and tables close to the text where you discuss them. If they disrupt the text because they are large, or there are many of them, then you may decide to group them all at the end of the section or the end of the report. Some organisations require that, although a lot of readers find it a nuisance.

All illustrations and tables must be numbered and must have meaningful captions or titles. And to state the obvious, they should never be there merely to look pretty or to fill a space. They are illustrations not decorations. Illustrations and tables are discussed further in Chapter 4 in Sections 4.10 and 4.11.

2.6 – Sequencing Level 1: The Conventional Format

We will now look at sequencing the report using the conventional format (there are other formats that we will look at later). There are three levels of sequencing to think about. The first is the obvious one and is what many people call the format, sections or structure of the report. The other two levels are sequencing the subsections within each section, and there could be many of them in a large report, and sequencing all the paragraphs within each subsection. This can all become quite complicated and you should keep it to the simplest level consistent with giving you a thorough plan that will make the flow logical and make the writing easier than it would otherwise be. We will first look in detail at a thorough approach to sequencing the report. Once that is understood we can then consider a simpler approach, which may sometimes be useful.

Let us now look at level 1, the conventional format for the sections. No format is right for every technical report but most reports have the same basic structure that follows the logical flow of the reasoning: state the problem, describe how you got the facts or data, present the facts or data, analyse them for meaning, draw conclusions from that meaning and then make recommendations about what to do about it all. The step-by-step flow allows readers to follow the argument from the beginning to the end – assuming they read it all.

This is the traditional approach. It is by far the most common format for technical reports and is easily adapted to both short and long reports. We will now outline the traditional sections and their sequence for both types. In Chapter 3, we will examine them all in detail. We will start with a typical sequence for a short report of, say, up to ten pages, although that length is not definitive.

Executive Summary
Introduction (Background and Aims)
Method – this could be the third part of the Introduction
Results
Analysis and Discussion
Conclusions
Recommendations.

Longer reports may need other sections and the Introduction may split into two or more parts that are sections in their own right. Here is a suggested full list for long reports but remember that few reports need every section listed here, not even all long ones of over, say, 50 pages. Only add extra sections if they are necessary.

I am going to use the terms front matter and end matter although they are usually applied to books. Strictly speaking, I am misusing them here but they are convenient for now.

Note: A Glossary can be in either the front matter or the end matter. Readers are more likely to see it, and therefore use it, if it is at the front than at the back, which is why I have put it in the front matter.

Front matter:
Executive Summary
Administrative details
Table of contents
List of figures
List of tables
Glossary.

Body:
Introduction/Background
Aims

Theory
Materials
Method
Results
Analysis
Discussion
Conclusions
Recommendations.

End matter:
Acknowledgements
References
Bibliography (rare)
Appendices.

Remember that you will probably never need every section in this list and your organisation's template should guide you as to which sections to use and their sequence.

2.7 – Sequencing Level 2: Subsections

Earlier, we mentioned three layers of sequencing: the major sections, the subsections within them and then the information within the subsections. We shall now look at the second layer of sequencing – the order of the subsections.

As technical reports get longer most of the major sections will split into several subsections and even those may have sub-subsections. Decide them all as best you can when planning. Even give them provisional titles if you can. Arrange them into the most logical sequence for your readers so that they collectively form a seamless whole – each moving naturally on to the next as viewed by the readers. Most of the time this will probably be in order of importance. Sometimes you need a different sequence such as a chronological order. You will not get this completely right the first time but keep at it until you are happy. Changing the order later is not a disaster but the earlier you get it right the better, even though in long reports you may need to do quite a lot of tweaking later.

You are building a storyboard. Always make your judgements with the intended readers in mind.

2.8 – Sequencing Level 3: Paragraphs

Once you have confirmed the main sections and their subsections it is time to move on to the third layer of sequencing, which is where the detailed planning takes place. This work will save you a lot of time when you start writing. There is no doubt that planning saves time and there is no doubt that some people will do more detailed planning than others. That is fine, and at least some of that is down to personality. What I am stressing here is that almost everyone will benefit from doing more detailed planning than they are used to.

Repeating myself, planning is mainly about choosing and sequencing the information for your report. This third level of sequencing pins down the order of the information within the subsections. Most people do this while writing and that is a rotten time to do it. Sequencing as much as you sensibly can before writing splits the tasks of sequencing the information and phrasing the information. Neither of those tasks is easy. Do them simultaneously and life will be more difficult than if you separate them. Remember the house building analogy – finalise the architect's plans before the builders start.

The following is a guide to help you to plan thoroughly so that when you start writing you can focus entirely on how to write, not what to write. Judge for yourself how much detail to include – and then go some way further. Adapt the depth of the process, the level of detail in the plan, to suit your personality. After a few reports you will find a happy medium that suits you and works well for you.

Start with bullet points

Do the third level of sequencing one subsection at a time. One relatively simple way to do this uses bullet points and there are two approaches to using them. (A different method uses mind maps or spider diagrams, which we will look at briefly later.)

The first bullet point approach produces a detailed plan. For many, this is the best way as you get the full plan before writing. The second bullet point approach produces a simple plan and you fill in the details when writing. This depends heavily on you knowing the material intimately, which you probably do, and just needing a trigger when writing. The simpler approach is quicker but may be less effective as it often means more rewriting later, which may mean you lose more time in rewriting than you thought you had saved through planning. Your personality will strongly influence your preference and I suggest you try both approaches a couple of times before confirming your choice. We will look at the detailed approach first.

To get a detailed plan, first make a bullet point list of as many of the points you can think of that you want to include in this subsection. You will have a lot of bullet points, which is why I suggest doing this one subsection at a time. Get them all in there and do not worry about the order or duplications; it may be messy, but you will sort it out later. In effect, this is a brain dump of what you want to include in this subsection.

Important: include your illustrations and tables, just as very brief titles for now. Also important: only use two or three words for each bullet point as they are simply labels or memory joggers for all the information you have in your head.

Group the bullets

When you are reasonably happy (it will be far from perfect yet) click and drag the individual bullet points into groups of bullets that belong with each other. Try to have only a few bullets in each group. If there appear to be too many in a group, then split the group in two. Make sure that all the bullets within a single group are related to each other. You are trying to get into one place all the bullets about one small part of your message.

Each of these groups is an embryo paragraph and, in effect, you are starting to create your paragraphs for that subsection. Include bullets of any illustrations and tables that belong to the group.

Sequence the groups

Now click and drag the groups into the best sequence for your readers, sequencing the embryo paragraphs for your subsection. The first group of bullets will become the opening paragraph of that subsection and should act as a scene setter or introduction. This scene-setter paragraph is sometimes called the topic paragraph and it often starts with a topic sentence – the attention grabber. However, do not become obsessed with topic paragraphs and topic sentences but do use them when they will help your readers.

Normally, sequence the rest of the groups in order of importance, although there are times when some other sequence, such as chronological order, may be more appropriate. Do this sequencing carefully so that you get a logical progression, which the readers will be able to follow.

Sequence the bullets

Now, group by group, sequence the bullets within each group. Click and drag them into the best sequence for that group. Where appropriate, the first bullet will become the topic sentence so try to include a key word or phrase that will focus the reader on what this paragraph is about. You are deciding the order for the things you will say in that paragraph, almost embryo sentence by embryo sentence.

Remember that the first group (the opening or topic paragraph of that subsection) should be an introduction to what follows. Sometimes (it does not always work) the sequence of the following groups will reflect the sequence of the bullets in the introductory group. That means that the order of the sentences in the introductory paragraph may set the order for the following paragraphs of the subsection. As I said, it does not always work out that way.

The last group of bullets will mark the end of the subsection and should contain the bulleted information that will bring it to a natural close.

Remember to keep all the bullet points short, just two or three words. As we said, they are your memory joggers; they do not need to make much sense to anyone else.

A simpler approach

For many people that detailed approach is the best approach but it will not suit everyone. A simpler approach is to write a single bullet point for each of the main points, each paragraph, you want to include in the subsection. Each bullet point is a label that must catch the essence of its intended paragraph. Then click and drag them into the order for the subsection. Collectively they should now outline the story of that subsection.

You may now decide to add some subordinate points to some or all of those single bullet points to make sure you do not forget something important later. The more bullet points you add, the more detail you add and the closer you edge towards the detailed approach we described above but probably settling for some middle ground between the two extremes.

Otherwise, your plan is ready and when you start writing you will add in the details for each paragraph guided by what is, in effect, the topic sentence for the paragraph.

Plan the emphasis

We often think of emphasis as being about bold, italic and underline but there is more to it than that and they may not be the best way to add emphasis to your text. A simple and powerful way to emphasise important points is to put them first. Start each subsection with a group of bullets that either states the most important point or points, or outlines the order of what follows. Start the other groups of bullets (embryo paragraphs) with the leading bullet (which will become the topic sentence). However, this is not painting by numbers so treat these suggestions as just that – suggestions, not rules – and ignore them when you feel you should.

Finalise your lists

Check through what is by now probably quite a long list. Change anything that needs changing, whether that is by adding or deleting bullet points, or by doing some rearranging. You may decide to keep some duplications if you have decided to repeat some important points but be wary of repeating too many points as that can irritate readers.

Few plans are perfect and when you come to write these out in full as real paragraphs you may want to reorder some items. That is fine; you are in charge.

Although this is time consuming, by taking it subsection by subsection you have sequenced all the points you want to make to your readers. They will appreciate the logical sequence, which makes the argument easier for them to follow, and you will save time when you start writing.

If you decide not to sequence at the planning stage, you will still need to do it eventually, usually when writing. Doing those two essential tasks (sequencing and phrasing) simultaneously is much harder than doing them separately and leads to more rewriting and more wated time. Writing is much easier when you already know what you want to say and the order in which to say it.

You now have a plan, a blueprint for your report which has sequenced sections, subsections, sub-subsections and even paragraphs. As mentioned earlier, put as much detail into this plan as you feel is sensible.

Here is a simple illustration of this bulleting process. It started with a fairly random brain dump, something like this:

- Bullet A
- Bullet C
- Bullet F
- Bullet E
- Bullet J
- Bullet B
- Bullet K
- Bullet I
- Bullet H
- Bullet D
- Bullet G.

Then it was arranged into three groups of bullets that belong together:

- Bullet F
- Bullet E
- Bullet G
- Bullet D
- Bullet H.

- Bullet A
- Bullet C
- Bullet B.

- Bullet K
- Bullet J
- Bullet I.

The three groups were then put into the best sequence:

- Bullet A
- Bullet C
- Bullet B.

- Bullet F
- Bullet E
- Bullet G
- Bullet D
- Bullet H.
-
- Bullet K
- Bullet J
- Bullet I.

And finally, the individual bullets were sequenced within their groups to produce three sets of bullets for three paragraphs, all now sequenced as the writer wants them to be:

- Bullet A
- Bullet B
- Bullet C.

- Bullet D
- Bullet E
- Bullet F
- Bullet G
- Bullet H.

- Bullet I
- Bullet J
- Bullet K.

How much detail?

How detailed you go with this technique will depend on several things including your personality. I strongly recommend that you try it and then modify it until you get an approach that suits you. Earlier, I said plan as much detail as you feel is sensible for you. In fact, the first few times you plan this way try putting in more detail than you feel is sensible. Only by going a bit too far to begin with can you find the approach that is best for you.

One tip: before you start writing show your bullet point plan to your line manager or whoever should sign off your report (it may need some explanation). Getting agreement to the plan at this stage can prevent wasting a lot of time later. Dealing with, 'I'd rather you did it this way,' at the planning stage is a lot easier than after you have written thousands of words.

I have even known occasions with short internal reports (never with external reports) where a detailed bullet-point list has been accepted as the report, making writing and editing a full report unnecessary. Now, that is a time saver. Of course, those bullet points need considerably more detail than two or three words can provide.

Plan section by section or plan the whole report?

One observation though, it is daunting to bullet point the entire report. You may find it much easier to bullet point one section at a time. Maybe you would then like to write that section before bullet pointing the next section. That is fine, whatever suits you. However, there is a slight risk of duplicating information between sections, but you can guard against that. If this approach suits you, then go with it.

Mind Maps

Instead of using bullet points, some people prefer to use mind maps to achieve the same purpose – to produce a design or blueprint that guides their writing. I can give a brief outline of the technique here but, if you are not familiar with using mind maps or spider diagrams, then look them up on the internet and try them.

Take a large sheet of paper, at least A3 in size, and write your report title in the middle. Then draw and label branches radiating off, like branches on a tree, one for each section of the report. Then draw and label smaller branches off the main ones, one for each subsection and so on. You may need different sheets of A3 paper for each section in order to include all the details. To replicate the three levels of sequencing described above, you will need branches (for the main report sections), sub-branches (for the subsections) and twigs and twiglets (for the embryo paragraphs and embryo sentences). Number them to record the sequence you decide on and remember to include labels for the illustrations and tables.

Essentially, this is the same process as for the bullet points but in an even more visual form. It has the advantage of showing a lot of the data all at the same time and allowing you to form links between the items with a pencil stroke. On the other hand, at some stage you will need to transfer it all into bullet points unless you can manage to write the draft text based on the branches, sub-branches and twigs – which can be done with a little practice. Personally, for my bigger writing projects, and for designing training courses, I sometimes combine the two approaches starting with detailed mind maps before turning parts of them into linear sets of bullet points – because the text is linear.

It really does not matter whether you prefer the bullet point approach, the mind map approach or a combination. Use whichever suits you the best. What does matter is having a detailed plan before you start the serious writing. How detailed you make it is up to you but try stretching yourself by getting a little more detail than initially feels right and see if it helps.

As we have said, do not try to write and plan simultaneously, which of course is what we tend to do naturally – it is a habit that needs to be broken. That habit makes the whole task a lot harder unless you are writing something that is very short, which you are not doing.

Once you have sequenced your plan, you have dealt with the logos part of the three rhetorical appeals and with the second of the three principles we outlined in the Chapter 1 (which were technically robust, arranged logically and written clearly). Now you are ready to write. The architect/designer has finished and the builder/writer can start. But, before we examine the tips on writing, we are first going to turn aside to think about section headings, then take a brief look at some other report formats available to us and then, in Chapter 3, work through a detailed description of all the typical sections of reports.

2.9 – Section Headings

At the planning stage, choose sensible headings for all the sections and subsections, even though you may change them later. The headings for all the main sections are likely to be the standard ones referred to above (Introduction, Method, Results, Analysis, etc.) or ones that are given in your organisation's template. You are more likely to choose the headings for the subsections yourself.

Headings are titles and they signal to the reader what to expect, so they must be meaningful. Readers will decide whether to read a section based largely on the heading along with, probably, the first sentence or maybe the first paragraph (hence the importance of those topic sentences and topic paragraphs). Pick titles that indicate the content of the section or subsection clearly; that is a service that all readers will appreciate whether they read or skip the section. You are not really trying to 'hook' the reader with your headings and opening statements; it is more a case of helping them to choose the parts of the report that matter most to them.

The headings also appear in the Table of Contents and many readers will use that to check what the report covers and see how it is organised, not just to see which page to turn to.

Heading styles

Different heading styles enable you to signal to the reader the logical hierarchy of sections and subsections. Use the 'Styles' in your word processor to give a consistent style to each different heading level. You can modify the default heading styles to suit yourself unless your template dictates them.

Most reports align all section headings to the left. You could choose to centre align Heading 1 but do not centre the others. The variations in style are usually self-evident and in order are a choice of capital or lower-case letters, font and font size, bold, italic and run-in. In addition, the line spacing above and below the heading can indicate its importance, with more space for major headings and less for minor headings.

How you use capital letters is a strong indicator of hierarchy. Heading 1 is usually the only one to use all capitals (TYPES OF BOILER). Heading 2 could have all the main words start with a capital letter, known as title style (Types of Boiler). Heading 3 and others could be sentence style – only the first letter of the first word being a capital (Types of boiler). The current trend is towards using sentence style for all heading levels, except perhaps for Heading 1.

The choice of font can be tricky although your organisation's templates may dictate which font to use. Traditionally, serif fonts (those with fancy bits on the ends of letters, like Times New Roman) are said to be easier to read than sans-serif fonts (without fancy bits, like Arial) although the letters of sans-serif fonts are thought to be clearer. While some argue the opposite, a glance at lengthy texts in books, newspapers and magazines shows that most of them use serif fonts. Take your cue from them and prefer a serif font if your report will be read in print. A sans-serif font may be better if most people will read it on a screen.

Heading 1 should have the largest font size. Subsequent headings can have decreasing font sizes until you reach the size used for the body text.

Bold usually signals a higher order of heading than italic. Underlining is little used, being seen as old-fashioned (it was about the only one available on typewriters and is now mainly used for web addresses).

Run-in headings. You will probably not need these. Run-in headings are for the least important headings. The run-in heading starts the first paragraph, on the same line as the first sentence, in bold with a full stop/period after it.

Numbering headings

Numbering sections and subsections is conventional in long technical reports but only number the first three or four levels. In fact, three levels are usually enough. If your template does not specify a numbering system, then try something like this:

1 Heading level 1
1.1 Heading level 2
1.1.1 Heading level 3
1.1.1.1 Heading level 4. (Although three levels are usually enough.)

Prefer not to progressively indent headings or you will march across the page and will also need to indent all the subsection paragraphs.

If you really must have more levels, then you can add them without appearing to do so by highlighting paragraphs using letters in round brackets/parentheses (a), then roman numerals in round brackets (i) and then ordinary bullet points.

Paragraph or line numbering

Although unusual in technical reports, numbering either the paragraphs or the lines of text is common in some organisations. It is especially useful where a group of people meets to discuss a long report as they can refer directly to a paragraph or line number. In MS Word, you can choose whether to number every line, every five or ten lines and so on, and whether to restart the numbering at every new page or section. You can also change the font setting.

2.10 – Other Formats

In this book, we are focussing on the traditional approach to formatting a technical report. However, there are several other ways to format technical reports and we shall look briefly at the main ones now.

The managerial format

Some technical reports follow a different sequence that states the problem or question and then immediately answers it before describing how the answer was obtained. The argument is that many readers will not read the main body (the heavy technical parts). They just want the answers. It is used when the main readers are managers, often senior managers, some of whom may not be qualified in the engineering discipline and will leave others to judge the technical information. The sequence follows the order of importance to these senior managers, not the logical flow of the argument.

Therefore, the sequence for a typical managerial format poses the question, answers it and then gives the details.

Executive Summary
Introduction
Conclusions
Recommendations
Method
Results
Analysis and Discussion.

Probably, you will only use this format if asked to do so by your manager or client.

The 1:3:25 format

This is another format for technical reports that is aimed at decision makers, although it is equally suitable for engineers. It was created by the Canadian Health Services Research Foundation (CHSRF) and was adopted by Britain's Health and Safety Executive for their research reports in 2003. In this format there are only three main sections to a report, although you can add appendices if needed. The numbers 1:3:25 refer to the expected number of pages for each section.

The 1: One page for the 'Main Messages' or 'Key Messages'. After a short introductory paragraph, the first section gives the main messages that decision makers need and is usually presented as a set of bullet points. The focus is on the conclusions, especially the consequences or implications of those conclusions. It is about what matters to the intended readers, what they need to know. It is not a summary of the methods or findings and may not even mention them. It answers the 'So what?' question. As with the executive summary in a conventional report, the Main Messages should be capable of standing alone without the rest of the report.

The 3: Three pages for the 'Executive Summary'. At about 10 % of the total, three pages for the executive summary is longer than in conventional reports. Use a sequence that starts with the information that matters most to decision makers and ends with the information that matters least to them: most important first. So, start with the issues you were examining, including the question or aims, and summarise your answers or conclusions, the recommendations and the consequences as appropriate. Limit the methods to about a couple of lines, they just need to know that you did the work properly, and briefly summarise the results. Non-technical managers should get all they need from the Main Messages and the Executive Summary.

The 25: Twenty-five pages for the main report. The CHSRF, the originators of the format, list seven subsections for the '25'.

They are:

- The Context, which includes the background and the problem or aims.
- The Implications for the decision makers (yes, it is still aimed at them), these are the Main Messages in their full glory.
- The Approach, which is a description of the methods used although anything really technical can go into an appendix.
- The Results, a summary only.
- Additional Resources, these point to other relevant information that decision makers are likely to find useful.
- Further Research, which points out knowledge gaps or questions still waiting to be answered.
- References or Bibliography.

The sequence of the main report of 25 pages or so is similar to the managerial format, giving the context, question and answer before any details about methods and results.

For most writers of technical reports, the 1:3:25 is an unusual approach but, like the managerial style, it is aimed mainly at decision makers rather than fellow engineers. Do not use it unless you have been asked to do so or have suggested it and gained agreement. Treat the numbers 1, 3 and 25 as forceful guidelines but not rules. Having said that, there is a variation for heavy technical or scientific reports that genuinely need far more space for detailed technical or scientific descriptions. This keeps the 1:3 but allows a lot more pages for the main report.

Email report format

Email reports (formerly memo reports) are short and informal. They are what they say they are, a report in an email. The informality stretches to using pronouns (which you would not use in a more formal report) so you can write, 'We did such and such...', 'I recommend that...' and so on.

Even though they are short and simple, they can still be important and may guide decision makers. Like all writing, they need a structure or format. Keep it simple, you may only need three or four paragraphs. The following two slightly different formats will guide you. The first is for a problem that you have investigated and fixed. The second is for suggesting a solution to a problem that you have investigated but not yet fixed.

For a problem you have investigated and fixed:

The problem

What caused it
The action taken
The result, how you have left it.

For a problem you have investigated but not yet fixed:

The problem
What is causing it
The proposed solution.

Although these are simple formats, they are widely applicable. The first is based on a well-known model: Situation, Task, Action, Result (known as STAR or, for some, Problem, Action, Result – PAR). The second is the equally well-known model: Problem, Cause, Solution.

Sales proposals

Sales proposals deserve a book to themselves, but I am not qualified to write it. They can be complex so we will not look at them in detail here, merely giving two tips. The first tip, if you are new to writing sales proposals, is to ask for guidance from someone who is both experienced and successful.

The second tip is that if you are replying to an invitation to tender, then read and follow the instructions very carefully as they often specify in minute detail how a bid must be arranged. Forget your own templates. Failure to follow the instructions in every detail may lead to your bid being rejected immediately.

This can be for two reasons. First, a common format for all bidders helps the evaluators to compare submissions fairly, even though it can be a bane for the writers. Second, this is the first task they have asked you to do and you ignored it. If you want to win, then do as they ask.

Other than that, the formats used for many sales proposals follow a similar approach with some variations.

Title
Purpose Statement
Executive Summary
Table of Contents
Introduction
Client's Problem – (our understanding of your problem)
Proposed Solution (our approach and how it meets your requirements)
Benefits to Client (not just the problem will go away but things will actually improve)
Timescales and Costs

Who we are
Appendices.

It is worth repeating, follow very carefully the instructions in the invitation to tender and be guided by someone who is experienced and successful in writing sales proposals.

2.11 – Chapter Summary

A good technical report meets several criteria, but it will not do this without careful thought and planning. These criteria start by asking and answering the question. They then include the three principles we outlined in Chapter 1 (technically robust, logically sequenced, and well written). They also include the three rhetorical appeals (ethos – the writer is credible, logos – the argument is convincing, and pathos – the readers are interested). There is some overlap and we can pull these together and expand them as follows.

First, it poses the right question and answers it, even if the question is not explicitly phrased as a question.

Second, the report is technically robust.

Third, it meets the needs of the intended readers and it allows the writer to add important points the reader was unaware of.

Fourth, it demonstrates the credibility and authority of the writer and his or her organisation to answer the question posed.

Fifth, its sequence of information progresses logically from the beginning to the end, and the intended readers can recognise that. It is comprehensive but not overburdened with unnecessary details, and it honestly reports contrary evidence and explains why it can be discounted, if it can.

Sixth, it is better and easier to do the sequencing before writing than while writing. Start with Level 1, the main sections, which are usually given in your organisation's template. Then move on to Level 2, the subsections, and Level 3, the paragraphs. Use bullet points for the main items of information and click and drag them until you get the sequence you want.

Seventh, the quality of the writing is such that it does not trouble the intended readers. The writing has a clear, flowing style and uses words that the readers understand. It enables the readers to understand it all at the first attempt – or gets as close to that as can be achieved.

3

STANDARD SECTIONS

The following detailed guide describes the content of every section of a technical report that I can think of, at least a report that follows the conventional format, from the Title to the Appendices. Very few reports will need all these sections and even fewer will need all the details given here. On the other hand, some may need other items that are special to that report.

It is worth repeating that this is a guide not a set of rules and you should follow your organisation's template if there is one. No format is right for every conceivable technical report so apply all of this with discretion.

3.1 – Title

Even the title has a purpose, which is to identify the report – and – to start to catch the readers' interest.

Catching the readers' interest may seem a curious thing to say but do not miss an opportunity to get potential readers to warm to your report. Scientific reports avoid this but there is some leeway with technical reports, but not much. Newspapers and magazines do it every day, of course, but do not go that far. Stupid or grandiose titles will fail your report before it is even opened and such titles have led to ridicule for reports that have entered the public domain.

While there is not a lot you can do to enliven your title, take some time to think about it and make it as meaningful as possible. You readers need to know what your report is about from the title and ideally it makes them think, 'I want to read this.' Do not worry about using too many words, within reason. You are fortunate that your likely readers have asked for your report and are already interested in it. Perhaps they are bursting with enthusiasm for it, or perhaps not. At least, do not bore them before they even start reading. If you need to persuade people to read your report, then a well-chosen title can help.

Occasionally, a subtitle may be appropriate. A subtitle gives you more freedom than the main title and can be an effective way to prompt interest.

Here are some examples of fictitious titles that are based on real reports; notice that they are informative but reasonably short. Potential readers will know immediately if the report is likely to be of interest to them.

Barnes Water Disposal Well, Review of Erosion Risk.
Renewing the Head Office IT Infrastructure.
Barnes Shopping Centre: Safety report on public-use lifts.
Report on the accident to Boeing 123ABC at Barnes International Airport on 9 May 2020.
Exposure to carcinogens in surface engineering.
Desk study of mineral reservations in Lancashire.

3.2 – Executive Summary

The executive summary is the most important section in the report. Almost everyone reads it or at least starts to read it, and many people read nothing else. It is essential.

The preferred heading is 'Executive Summary' although some organisations use 'Management Summary' or 'Summary'. Never use the single word 'Summary' if there are summaries at the end of sections.

The executive summary should give a short, self-contained overview of the entire report – who asked for it and when and why, the aims, a very brief description of the main work, and a summary of the main conclusions and recommendations. Basically, it gives the question and the answer. See it as the only source of information for someone who will read nothing else but still needs to be accurately informed, although it also serves as a brief overview for those who will read much more.

It must be self-contained because, even when it does not have to, it must be able to stand alone without reference to the rest of the report. We know that some people will read nothing else and some may not even have access to the rest of the report. So, it must function without support and must not make cross-references to information, figures or tables in the report (such as see section 2.5 or see Figure 3.2). In this sense, think of it as a separate document and not part of the report. (Regard appendices in the same way.)

If you number your main sections, then the executive summary will not have a number (just as appendices do not have numbers, although they do have letters). Likewise, it is likely to have an italic page number rather than a roman one. Now these two points are a bit pedantic and many writers ignore them, making the executive summary Section 1 and placing it on page 1, not page i. Check your template or other reports from your organisation, otherwise decide for yourself.

Traditionally, it appears on the first page after the cover although nowadays it sometimes comes after the administrative details but usually before the table of contents.

State briefly the gist of the report, capturing its essence. If someone reads nothing else, what must, should or could they know? They must know the aims of the report (the question) and its background so that they understand its relevance and importance – why it was needed and probably who asked for it. They must know the main conclusions and recommendations (the answer). They should know enough about the methods to believe that the work was thorough and professional. They could be told a little about the main results, probably focussing on the scale rather than the results themselves, with just enough detail to convince them that your conclusions are based on an adequate amount of data.

If you have a large report with many conclusions and recommendations, then either include them all or give the most important ones while clearly stating how many other, relatively minor ones, are in the main report. If the readers have not seen them all, then they must know how many they have not seen, their topic or subject areas and that they are less important than the ones they have seen.

Never cut and paste from the main report. Remember that, essentially, the executive summary is a separate document and more people will read it than any other part of the report. Cut and paste rarely works well because the words and sentences were written to fit smoothly somewhere else. Do the executive summary justice and write it properly.

The executive summary can be the hardest part of a report to write. Write it last. It is not a dummy run for the main event; for some it is the main event. Wait until you have the full report before you try to summarise it.

Use plain English throughout the executive summary as you will have non-specialist, non-technical readers for this section. Plain English does not mean simple English, you are not writing for children. It means using words your intended readers will understand, which will mean cutting out as much jargon and acronyms as you can. You cannot always find ordinary words to replace jargon words, which is why we use jargon in the first place. Where it is impossible to avoid using jargon, take extra care to explain what you mean as clearly as you can for someone who is not a specialist. It is a good idea to avoid all acronyms in an executive summary, spelling each term out in full. Some organisations take this literally, even banning the terms UK, EU and USA.

Figures and tables are rare in executive summaries, but you can use them if they are the best way to put information across to a wide range of readers. However, take care not to give more information and detail than your wide range of readers need at this stage.

You will probably use the past tense most of the time, but you will need the present tense for the aims, conclusions and recommendations (the aims are…, the conclusions are…, and the recommendations are…).

Here are two models that might be useful when planning some executive summaries, but they will not work for all. We mentioned them earlier when we looked briefly at email report formats.

The first is: Problem – Cause – Solution. This describes or explains the problem, says what is causing it and tells how to fix it.

The second is: Problem – Cause – Action – Result. This describes or explains the problem, says what caused it, tells what you have done about it and then describes the results of those actions. (The STAR model is a better known version of this: Situation, Task, Action, Result.)

3.3 – Administrative Section

Your organisation will require some administrative details and your template should define what is required and where it must appear. If not, look at other reports from your organisation to see what has been approved in the past. Very short reports may require few details and those details can sometimes be placed on the title page.

You may need any of the following:

- Reference or document number, version or revision status.
- Client or customer name and identifier or contract number.
- Author's details and department.
- Checked by, approved by, authorised by.
- Dates of work, from, to.
- Date of issue.
- Confidentiality or security classification.
- Distribution or circulation list.
- Key words if needed for meta data to trigger search engines.

3.4 – Table of Contents

Many readers scan this to get an overall impression of the report. Include all major sections and appendices, with page numbers. Many tables of contents only show the main sections of chapters and that is usually sufficient. However, consider including the Heading 2 subsections if you think that would help your readers. You may not need a table of contents in reports of only a few pages.

Remember that sections have numbers whereas appendices have letters.

3.5 – Lists of Figures and Tables

Few reports require separate lists of figures or tables but it may be sensible to consider them, albeit briefly. They may be used in long reports but only if you have a lot of figures or tables. The list of figures should come first. In short reports, or where there are few figures or tables, such lists can appear fussy.

3.6 – Glossary

A glossary is a list of the terms, acronyms and abbreviations used in the report. Many readers will find it helpful, including technical managers who may be somewhat removed from the latest technology and its terminology. Of course, you must still define all abbreviations in the report the first time you use them – but do not expect everyone to remember them. You will probably introduce most acronyms in the heavy technical parts of the report which few non-specialists will read. If you then reuse those acronyms in the conclusions and recommendations, many of the non-specialists who will read those sections (and see them as very important) may not understand them. Redefine them as an act of courtesy to those readers.

The glossary traditionally appears towards the end of the report, just before the acknowledgements and references, but consider placing it near the front as readers are more likely to notice it there, and use it, than if it is at the back.

3.7 – Introduction

The introduction is an essential section. Its job is to set the scene for the report, and what is in it depends on the length and complexity of the report. Three questions must be answered: 'Why' – the background, 'What' – the aims, and 'How' – the methods used. Why was it necessary to do this work? What was it meant to achieve? How did you do it? In short reports, cover all three in the introduction. In long reports, the methods have their own separate section leaving the introduction with just the background and aims.

Therefore, there are two common approaches. First, the introduction covers all three parts – the background, the aims and the methods – probably with the methods being in their own subsection with their own subheading. Second, the introduction covers the background and the aims, but the methods, which are now more complicated to describe, are moved to become a main section on their own.

Whether to separate the methods description is something that the writer must judge, probably guided by his or her organisation. Different people and organisations will make different decisions. In many respects it does not matter provided the readers understand all they need to know about all three areas.

All of your readers need to understand the aims of the report and most of them need to understand the background. Therefore, keep the technical language to its simplest level and avoid jargon and acronyms if you can.

We will now look at each of these parts in turn, although to save repetition we will leave the details of the methods until we discuss them as a separate section.

Background

The background states the problem and its significance, giving the reason and motivation for doing this work and your mandate for doing it. It answers the 'Why' question we referred to earlier. Answering the 'Why' question is vital if your report enters the public domain. Such reports can have a wide range of people reading them, from individuals with a vested interest to local and national media looking for a story. Why did you do this work? Was it reasonable to spend all that time and money, maybe public money, on this work? Can you justify it all?

The background may also answer the 'When' and 'Who' questions. Give the start and end dates of the work and any other important dates. For some reports, you may be expected to name the people or organisations who were involved in different parts of the background and those who did the work itself. That could be part of the ethos rhetorical appeal.

Sometimes, the background may summarise earlier thoughts or work that are relevant to your report. It is rare for technical reports to need a complete literature survey but those that do should present it as a separate section.

Finally, is there any chance that your report may be pulled from the archives and referred to in several years' time? This can happen. For example, geologists can refer to reports that were written decades earlier. If there is a chance of this, then it is wise to write a detailed background as an appendix to your report. Today's readers may not need all that detail so it should not be in the introduction, but readers in the future might appreciate it. Just think, some of your readers may not have been born yet.

Most of your readers will want to understand the background so keep the technical language to a minimum, and probably use the past tense for most of it.

Aims/Purpose/Terms of Reference/Scope

The aims are, of course, vital and, as we have already noted, they are often given their own subsection and subheading in the introduction. Many people use the terms aims, purpose, objectives, terms of reference and scope quite loosely to mean much the same thing. In many reports that is not much of a problem, but there are differences.

The aims could also be called the purpose and those terms are more or less interchangeable. Sometimes they are called objectives, which is a bit of a misuse of the term. All three are 'What' questions. There are no standard definitions for these terms, but the aims or purpose are usually seen as the overall or high-level target or targets, whereas objectives are specific steps along the way to achieving the aims.

For example, a football manager may aim to win the league in the next two years but that could include objectives to change the coaching staff this year and buy two new players next season.

It is usual to write aims beginning with the word *to*. In other words, to use the infinitive of the verb as in 'to win' above.

As already noted, the aims can often be phrased effectively as a question or several questions. (See Chapter 2.)

The scope of a report defines the boundaries of the investigation, saying what is included and what is not included. This could apply to time limits, geographical limits and so on. Sometimes, the interaction between time, costs and quality can set boundaries to a report.

Terms of reference is the widest-ranging term and will include the aims, objectives and scope of the report. The terms of reference are often set by the person who requests the report, with or without discussing them with those who are doing the work and writing the report. The terms of reference may also name the personnel involved and their responsibilities, and the financial and time constraints.

Note that aims are usually written in the present tense, using plain English as much as possible as every reader must be able to understand the aims easily.

Depending on the report, placing a simple statement of the aims in the introduction may be all that is needed, but as things get more complicated you may decide to add a scope statement or eventually a full terms of reference. Whichever you use, it is sensible to define these things before you start the investigation, or at least before you start planning the report. However, realistically, there are occasions when the aims can be a bit fuzzy in the initial stages of an investigation but do clarify them as soon as possible. With a report, you will always want the aims and scope (if applicable) to be clear the moment you start planning – otherwise you are heading for a rewrite after being told, 'It's not quite what we wanted.'

Very unusually, like the methods, the aims can break away to become a separate section leaving just the background in the introduction. This is rarely necessary and usually only happens in some long and complicated reports.

Methods

We cover these in more detail below where they would appear if you wrote them as a separate section.

Obviously, the methods (or procedures or techniques) tell how you did the work, especially how you obtained the data – answering the 'How' question. When they are part of the introduction, you need to describe them to readers, including non-technical people, in such a way that they believe you did the investigation in a professional and competent manner. This means cutting out jargon as far as possible. Many of them will neither want nor need the details. Consider moving the methods to a separate section if you have readers who do need a lot of details.

The huge unwritten objective of describing your methods in commercial technical reports is to establish your credibility (ethos) in the eyes of your readers, telling them enough to prove that you did a thorough and professional job, while not telling them how to do it for themselves.

3.8 – Theory

A theory section is rare in technical reports but common in scientific research reports. You are not writing an academic thesis so only include a theory section if you feel your specialist readers need it to help them understand the results and analysis. Minor bits of theory could be included, or cited, in the analysis section if necessary. If you really need a theory section, then you may also need a literature review. However, both are well beyond the scope of most technical reports.

3.9 – Materials and Equipment

In some technical reports the readers will need to know about the materials you have used in your work, chemical substances for instance. If necessary, include a section describing those materials, such as how they were obtained, their purity or contamination levels and storage facilities. Include anything that an informed reader may want to be reassured about.

In a few technical reports, you may need to list any specialist test equipment you used, possibly with the calibration dates. However, that is more likely to be needed in science reports.

3.10 – Method

As already noted, describing your methods or procedure is essential and for most technical reports it will be covered fully in the introduction. In lengthy reports, the methods may become more complicated and you may decide to move them to a separate section, which is what we are now considering.

If you have not included the methods in the introduction, some of your non-technical readers will look here to get an idea of how you did the work. As they may not be able to follow a complicated technical explanation, start with two or three paragraphs aimed at them. Briefly outline your methods in plain English in just enough detail to convince them that you did a professional and competent job. This is part of the ethos rhetorical appeal so give them just enough so that they can trust you. Then you can move on to the technical readers.

Your technical readers will take much more convincing and will need considerably more detail before they are willing to give their nod of approval to their non-technical managers. Your hidden aim is to reinforce your credibility with these specialists who will understand exactly what you did and who are probably reading it on behalf of senior managers as well as for themselves.

The methods section presumably originates from the procedures given in scientific research, but there is a huge difference. In scientific reports the writer provides enough detail to enable other scientists to repeat and check the work. It is unlikely you will want to do that in your reports as you would like your clients to come back to you with further work at some time, not do it themselves.

The detail needed varies enormously but could include some of the following:

- The methodology or overall strategy, including any internal or external standards used if any (such as from the International Organisation for Standardisation (ISO), the British Standards Institution (BSI), American National Standards Institute (ANSI), your own Standard Operating Procedures (SOP) and quality procedures).
- The methods or techniques used to gather specific information or data.
- The traceability of samples or materials, if applicable.
- The test instruments and calibration status.
- The processes taken to ensure reliability, avoid contamination, verify results, etc.
- The constraints, accuracy, resolution, tolerances, etc.
- The assumptions: what they are and how you justify making them – although some of that may be more appropriate in the analysis section if there is one.
- The time scales, if pertinent.

You may decide to include very briefly how you analysed the data, although the full description will come later in the analysis section.

You are most likely to use the past tense when describing your methods.

Methodology

There is a difference between methods and methodology. Methods are the procedures or techniques you used or, in other words, what you actually did. This is the term you will normally use.

Methodology is not a posh word for methods. Methodology refers to your general strategy or approach, including why you chose the methods you used and not others, and why the chosen methods should give meaningful results. Methodology can also mean a recognised system or set of methods, techniques or procedures for doing something. Such a set of methods may be defined by your internal quality procedures or by an outside body such as a standards institute. Methodology can also mean the study of methods.

Normally in short technical reports you will describe your methods or techniques, not your methodology. In long reports you may indeed be describing your methodology, especially if you have to describe your general approach or strategy as well as the specific techniques you used.

3.11 – Results or Findings

This is usually a major section in any technical report and is likely to be split into subsections, all arranged logically and each with its own heading. Together, they present the findings your readers need to understand how you reached your conclusions. At this stage, almost all your readers will be technically qualified.

Some of them will be interested in seeing more information than you have included. You will have left it out because they do not need it to understand how you reached your conclusions – and that is the right decision as it helps to keep the main report short. Such additional information should go into an appendix. Take care though that all the information the readers do need to work out the conclusions for themselves is in the main report – not shunted into an appendix.

As you are now talking to fellow specialists you can freely use the appropriate technical jargon and acronyms. However, stick to the old rule by defining acronyms on first use, even though your readers should understand them anyway. As mentioned before, if later you use them in the conclusions or recommendations then redefine them there as you will pick up new readers.

Present your findings in a logical sequence following the plan you designed earlier, which may be in order of importance or chronology or any other sequence you feel is best for the readers. You will have already chosen a sequence you think will work well. Remember that this is the results section and simply presents your results or findings; there should be no discussion or opinions here. They come later.

Include all the details needed to make your case. Together, these facts must fully support your conclusions. Many readers will want to check that your facts lead logically to your conclusions and some may derive the conclusions themselves from what you tell them here. Include any findings that seem to conflict with your conclusions. Do not hide them or wish them away; they are part of your findings and your readers ought to be aware if there are possible conflicts. Later, in the discussion section, you can explain why you discounted them.

Use the illustrations or tables you planned earlier to help your readers to grasp complex data, information or issues and give each of them a reference number and caption or title.

Use the past tense to state your results: 'The trigger voltage was…', 'Annealing began at…', 'The liquid had evaporated…' Use the present tense when referring to figures and tables: 'This is shown in Figure 2.' or 'See Table 6.'

Independent sets of results

In large reports, there may be sets of results from independent topics all of which you bring together for discussion later in the report. In some reports there may be dozens or even hundreds of separate topics.

For example, in a report into an aircraft accident every component in the engine and fuel system might have been tested with each having its own independent test procedure and results. In such circumstances, it would be silly to describe, let us say, 20 procedures followed by 20 sets of results. Instead, pair the methods and results for each topic, item or component.

Such an arrangement might look like this:

Item 1 – method 1 and results 1.
Item 2 – method 2 and results 2.
Item 3 – method 3 and results 3, etc.

As always, think about your readers and try to see things from their perspective.

3.12 – Analysis

Your purpose now is to draw out the meaning from your raw results and present it to your readers clearly, but it is not yet discussion time. Collectively, what do these results, findings or facts mean?

In many technical reports, the results by themselves may not mean much to many of your readers unless they apply the same amount of effort to understanding them as you have already done. Your job now is to save them the trouble of doing that by leading them by the hand.

Extract the meaning, find the trends, fit the data to equations and make extrapolations or whatever. Lead your readers forward but save any discussion of the impact this new knowledge may have on the readers until the discussion. This is the process of turning raw data or measurements into useful knowledge and it is a crucial part of most technical reports. The discussion section will take this further by examining consequences and implications, and giving considered opinions.

Sometimes the analysis may be complicated. The specialists will be interested and will want to understand your reasoning. Many non-specialists will probably avoid it, especially if they just want the answers. Even the title may put them off.

The analysis should never contain any new results or data, although it should derive new information from the results or data – the meaning that the data is revealing. That meaning is important to the readers. Think of the data sent to Earth from a distant spacecraft. That mass of ones and zeros means nothing to most of us, but we are filled with wonder when we see the resulting photograph of a pale blue dot that is the Earth viewed through the rings of Saturn.

In short reports you may decide to combine the results and analysis into a single section as a natural way of putting the two together. Alternatively, you may prefer to combine the analysis with the discussion, which normally would be the next section. However, keep all three separate in long reports.

Mostly, you will use the present tense in the analysis such as 'The observed results obey the inverse-square law.'

3.13 – Discussion

In the discussion, you take the meaning that the analysis revealed, relate it to the aims of the report and interpret it for your readers. There will be a discussion section in most technical reports and it is likely to be crucial to some of your readers. Many of them will have skipped the methods, results and analysis but will now start reading. They do not want to know (and may not understand) exactly what you did and what you found – but they do want to know what it means for them. They want to know the implications and that may include the costs.

Here you show your mettle. Just possibly a non-specialist could have obtained the results or measurements, it is less likely that they could have done the analysis, and it is highly unlikely that they could give any meaningful comment as to what it means. Here in the discussion, you give and explain your expert technical opinions. It is worth repeating that your readers want to know what the implications are – for them – of all that you have found. If there are options to discuss, dispassionately explore in detail all the advantages and disadvantages of each.

At times it will be natural for some options and conclusions to emerge as you write. Let them do so, it is normal for options and conclusions to emerge during the discussion. This is not the latest blockbuster novel so do not leave your readers on a cliff edge waiting for the next chapter. There is this, this and this, therefore… conclusion. The next section will list all the conclusions in one place but let them emerge here if they do so naturally. Some of your readers will reach the same conclusions as you did while they are reading the discussion.

As options emerge you will want to discuss their relative merits and that may include relative quality and cost issues. Summarise these in the conclusions section and offer a decision in the recommendations. In some lengthy reports it may be better to split the discussion into a technical part followed by a financial part, possibly giving the financial part a separate section of its own.

Also in the discussion section, you will address and try to resolve any contentious issues or awkward findings that conflict with your conclusions. If you think that further work could resolve contradictory findings, then say so but do not specifically recommend it. Recommend it later – in the recommendations. If you are dismissing contradictory or conflicting findings, then give your rationale for doing so. Do it persuasively and convincingly, so that your readers should accept your reasoning.

If they lie within the scope of your report, discuss any wider or longer-term issues that your analysis reveals. If they are not within the scope of your report, then simply mention them and say that they lie outside the scope of this report. If appropriate, in the recommendations you might suggest further work to widen the scope.

Write the discussion in the present tense (as if it were a discussion here and now with your readers) and use plain English as much as possible, avoiding acronyms and jargon if you can or explaining them as you go along. Remember that new readers have joined you and they are not specialists so define any acronyms you feel you have to use; almost certainly they will not have seen any previous definitions.

Above all, remember that these new readers probably see this as one of the most important parts of the report.

3.14 – Finance

Although technical reports are about technology, many also need some information about finance. After all, any actions you recommend are likely to cost money to implement. Where to include the financial information can be a dilemma.

You will probably want to settle all the technical aspects of your report before considering costs, or costs and benefits, in which case you may choose to leave the financial considerations until the discussion section because they are a consequence of the technical aspects. You may find that they fit naturally towards the end of the discussion. As mentioned above, in lengthy reports they may even merit an entirely separate section after the technical discussion. However, as finance like all the other issues in the report has two aspects to it (facts and your professional opinions) you may decide to deal with the factual aspects in the findings and discuss their consequences along with everything else in the general discussion. As usual, have a look at how other similar reports in your organisation have dealt with finance.

The aims or purpose of the report, agreed with the client at the outset, may say if a separate section is needed and, if so, what financial information is required. If you need a separate finance section, then seek advice from your manager or your department's management accountant. If you are lucky, they might draft it for you.

Regardless of where you discuss the financial considerations, you may have to explore the relative costs of the alternatives you are discussing and the financial restraints that might limit the choice. Once you are discussing costs, you may have to consider the relative quality of the options – whether they are gold, silver or bronze solutions. Time is the third element; the better the quality of the solution the more likely it is to cost more, but it might last longer or take longer to implement. This is sometimes called the time-cost-quality triangle; changing one usually affects one or both of the others.

Which option is best when you consider the effectiveness, costs and lifetime of the solution? Is it better to buy a new and expensive piece of equipment or to extend the life of the existing one whilst knowing that obtaining spares will become ever more difficult? This is an old and common dilemma. Should we replace the IT system now or keep patching it for another year or two? Should we buy or lease a new crane or struggle on repairing the old one? Should you keep fixing your old car or get a new one?

3.15 – Conclusions

Every technical report should have some conclusions – firm conclusions, the time for discussion is over and here is the judgement. Just as in a trial in a court, the prosecution and defence have made their case and now it is time for the jury to give their conclusion, short and snappy. But there is one big difference between a jury's verdict and your conclusions; here, there are no surprises. Just as in a court case, there is no new evidence either.

Your conclusions are factual statements based on the results you have obtained and your deliberations as to what they mean. Conclusions do not make suggestions (recommendations) about what to do as a result. Qualified people should be able to derive your conclusions from the information you have already given them in the preceding sections, just as you did. Indeed, they may already know some of your conclusions because they emerged naturally in the discussion section. I repeat because it is important: no surprises and no new evidence in the conclusions.

Use plain English as far as possible as many non-technical readers will read this section and must be able to understand it – they may be the ones who will eventually have to decide whether to act on your recommendations, which will be in the next section of the report. For their sake, try to avoid jargon and acronyms where possible. As with the discussion, if you do use acronyms then redefine them here for new readers.

Each conclusion is a separate piece of information so give each one its own paragraph. In short reports where short statements will suffice you may consider using bullet points or a numbered list. Bear in mind that numbering a list clearly suggests a sequence that many people will assume to be an order of priority, whereas they are less likely to make that assumption with bullets.

Whatever the length of the report, make sure that each conclusion is clear, concise and self-contained, although they can contain carefully explained caveats if necessary. Above all, every reader must be able to understand them – and it is your job to make sure they can. When you have several conclusions, consider saying how many there are before you start to describe them.

Think carefully about the order in which you will present them. The order in which you reached your conclusions may not be the best order for your readers to understand them, especially for those readers who have not read the main body of the report. Often giving them in order of importance is a good choice although a chronological sequence may be more sensible for some reports. No single sequence suits all reports.

Normally, you do not need to justify or explain your conclusions – all the justification should have been in the discussion. Use bold factual statements, not hesitant or tentative statements. However, you may have some conclusions that are tentative or issues that are unresolved and need further work to clarify them. If so, make that clear.

If you have concluded that there are several options for some issues, then summarise the advantages and disadvantages of each option, but do not make any recommendations yet.

For unresolved issues, say either that they require further work or they can be ignored, perhaps with provisos.

Use the present verb tense for all or nearly all the conclusions, although you may also need the past tense in places. It is as if you are talking to the reader here and now: 'I have concluded that this is the case, although it was different before the repairs were completed.'

3.16 – Recommendations

Finally, it is time to recommend what the reader should do to achieve the original purpose of the investigation. Perhaps that is to put things right where they have gone wrong, perhaps it is how to make a site safe for use, perhaps it is to change a road junction because there is now more traffic than when it was first built, or whatever. This is the equivalent of a court judge passing sentence or releasing the accused if found not guilty.

However, unlike court cases, recommendations are not needed in all reports. For example, some consultants are hired to investigate a situation and draw conclusions, but the client reserves the right to decide what to do about it all. Be clear at the beginning as to whether recommendations are expected or not; this should be clear in the aims or the terms of reference. Usually, your readers definitely want your expert opinion and will expect recommendations. Whether the readers act on your recommendations is, of course, up to them.

As with the conclusions, each recommendation is a separate piece of information so give each one its own paragraph or, in short reports, consider using bullet points or a numbered list but remember that numbers suggest a sequence or priority. Whatever the length of the report, try to keep each recommendation short and to the point, without embellishment. Every reader must understand them – and again, it is your job to make sure they can do so. As with the conclusions, you may choose to introduce them by saying how many there are.

Use plain English as much as you can because non-technical readers will read this section and they must be able to understand it fully; they may be the ones who will decide whether to carry out your recommendations. Be bold, avoiding hesitant or tentative statements. If you need to add a caveat, then do so but remember that your readers are looking for answers not more questions.

Consider using 'so that' statements if you think they will help the readers. 'Do X so that Y will happen,' may be more helpful than a simple, 'Do X.'

Choose a sensible sequence, based on importance, urgency, chronology or whatever. If each recommendation links directly to a single conclusion, then use the same sequence in both sections or consider combining them in a single section (see below).

If you want to stress the relative importance of the recommendations, you could use the trigger words must, should and could. This must be done, this should be done, this could be done and so on. Sometimes you will recommend not doing something: this should not be done. In some business circles this is referred to as: MoSCoW – must, should, could, won't. (Who dreams up these acronyms? Maybe you could use it in a pub quiz.) Alternatively, you could use a numbered list.

Combined conclusions and recommendations

Sometimes in reports, every recommendation derives directly from one, and only one, conclusion and each pair is independent of the others. In such cases, consider merging the conclusions and recommendations sections into a single section. This will probably be more helpful to the readers than keeping them separate. In this way you could present Conclusion 1 and Recommendation 1 together, then give Conclusion 2 and Recommendation 2, and so on.

Do not even think about doing this if the recommendations are based on several conclusions.

3.17 – Acknowledgements

If you work for an organisation, especially a consultancy, it is likely that your senior managers or partners will decide whether you should include an acknowledgements section in your report. There will probably be a standard, stand-alone paragraph to which you can add the necessary details. If not, find out if other similar reports from your organisation have included acknowledgements and take your cue from them. Ethics is important in engineering and if others have contributed in some way to your work then their contribution ought to be acknowledged.

3.18 – References

Many technical reports need to point to supporting documents. Obviously, references enable readers to check original documents if they want to, but references can also reinforce your credibility or professionalism and lend support to the ethos of the report. In a sense, they say you are confident about your work and you are happy for readers to check things for themselves.

The terminology of referencing can be confusing. A reference is the article or book you are referring to and is listed with other referenced works at the end of the report but before any appendices. A citation is the place in your text where you point your reader towards the reference. The confusion arises when they are both referred to as references, which they are not.

Methods for citing and listing references vary considerably between publishers, editors, authors and academic disciplines and there are several styles to choose from. However, for technical reports the choice is between two widely used styles: the Vancouver or number style, and the Harvard or name style. Both have been around for over a century. Each has a standard format that should be followed but the minor details vary a lot, particularly the punctuation. What really matters is that readers understand each reference easily and that you are consistent in how you use your chosen style. One common, but not universal, principle is that you write the title of a report or book in italics so that it stands out. For an article published in a journal, it is the journal's title that is in italics, not the article.

Both styles are valid for technical reports and both are widely used in scientific and engineering journals. Find out which style your organisation or your client prefers.

In both styles, the essential information is the author's name (in some instances it will be the editor's name), the title, the year of publication and the publisher. Include the place of publication if you can find it.

According to Wikipedia, although the Vancouver system has been around for over a hundred years, it only acquired its name in 1978 when a group of editors of medical journals met in Vancouver to agree a set of standards for use in medical journals. Also according to Wikipedia, the Harvard style was first used by a professor of anatomy at Harvard University in 1881 and nicknamed the Harvard system by a visitor.

Vancouver/Number system

In the Vancouver/number system the references are numbered sequentially in the order in which you cite them in the report. So, reference 1 is the first to be cited, reference 2 the next and so on. Each reference has a unique number which you use every time you cite it. List the references in numerical order in the reference section at the end of the report. Prefer to put a full stop/period after the number (as shown below). Name all the authors if there is more than one, and note that a comma is not needed between the author's surname/family name and his or her initial(s).

All references are cited in the text as a superscript numeral[1] or as (Ref 1) or (1) or [1]. Preferably place the citation at the end of the relevant sentence unless that would be ambiguous.

One potential disadvantage of the Vancouver system is that you may have to make a lot of corrections if you decide to add or delete citations after writing the report, and then have to renumber many of them.

Vancouver – general format

Books:
Author. *Title*. Edition. Place: publisher, year.

Journals:
Author. Article title. *Journal title*, year, volume number (issue number), page numbers.

Reports:
Author. *Title*. Organisation. Report identifier. Year.

Standards:
Name of the standards institution. Standard number. *Title*. Place: publisher, year.

Online publications:
Follow the appropriate format for the type of publication and then add: 'Available from: URL. [Accessed date]'. Give the date in an unambiguous style: 3 August 2020 or August 3, 2020. Or use the international standard: 2020-08-03.

For example:

1. Austen J. *Modern Polymers*. 2nd ed. London: Wigchurch Press, 2017.
2. Wells H. *Polymers*. London: Solidbrick Press, 2020.
3. International Organisation for Standardisation. ISO Standard No. 45001:2018. *Occupational health and safety management systems – Requirements with guidance for use*. 2018.
4. Austen J. *Polymers and their applications*. London: Wigchurch Press, 2020. Available from http://www.xxx. [Accessed 27 July 2020]
5. Dickens C. The Tay Bridge Disaster. *British J Construction*, 2019, 31 (2), 63-71.
6. Green D. *Land slip risk at Site XYZ123*. ABC Consulting. Report Number: 2018/S196. 2018.
7. Dickens C. The Tay Bridge Disaster: The aftermath. *British J Construction*, 2020, 32 (5), 213-228. Available from http://www.xxx. [Accessed 3 August 2020]

Harvard/Name system

The Harvard/name system identifies references by the author's (or editor's) surname/family name and date of publication. Citations in the text are given as the author's surname/family name (no initial) and year of publication, all in one set of round brackets: (Austen 2017). If you are listing more than one source published in the same year by the same author, then label them as 'a', 'b', etc.: (Austen 2017a), (Austen 2017b).

The references at the end of the report are listed in alphabetical order using the main author's surname/family name and initial followed by the year of publication. A comma is not needed between the author's surname/family name and his or her initial(s). If the same author has more than one reference listed, then his or her multiple entries are listed in date order. If he or she has more than one publication in the same year, then list them as 'Austen J. 2017a' and 'Austen J. 2017b', etc. If there are several authors, then give all their names. The year of publication can be placed in round brackets if you prefer.

One potential disadvantage of the Harvard system is that the citations can occur in the middle of a sentence and that may disturb, in a minor way, the flow of the sentence.

Harvard – general format

Books:
Author. Year. *Title*. Edition. Place: publisher.

Journals:
Author. Year. Article title. *Journal title*, volume (issue number), page numbers.

Reports – based on that for books:
Author. Year. *Title*. Organisation. Report identifier.

Standards:
Name of the standards institution. Year. *Title*. Standard number. Place: publisher.

Online publications:
Follow the appropriate format for the type of publication and then add: 'Available from: URL. [Accessed date]'. Give the date in an unambiguous style: 3 August 2020 or August 3, 2020. Or use the international standard: 2020-08-03.

For example:

Austen J. 2017. *Modern Polymers*. 2nd ed. London: Wigchurch Press.
Austen J. 2020. *Polymers and their applications*. London: Wigchurch Press. Available from http://www.xxx. [Accessed 27 July 2020]
Dickens C. 2019. The Tay Bridge Disaster. *British J Construction*, 31 (2), 63-71.
Dickens C. 2020. The Tay Bridge Disaster: The aftermath. *British J Construction*, 32 (5), 213-228. Available from http://www.xxx. [Accessed 20 July 2020]
Green D. 2018. *Land slip risk at Site XYZ123*. ABC Consulting. Report Number: 2018/S196.
International Organisation for Standardisation. 2018. *Occupational health and safety management systems – Requirements with guidance for use*. ISO Standard No. 45001:2018.
Wells H. 2020. *Polymers*. London: Solidbrick Press.

As already mentioned, neither system has a fixed style for punctuation. However, there is a general trend in business writing to simplify punctuation, for instance writing 'Mr J Smith' rather than 'Mr. J. Smith'. Follow your style guide if you have one, otherwise make your choice and apply it consistently.

Bibliography

It is highly unlikely you will need a bibliography (or list of further reading) in a technical report as you will have given references to any documents you have cited and will rarely need to suggest anything else. However, one thing a bibliography could be used for would be to list documents you have consulted but not cited. These could, for example, include other reports or standards publications. However, for simplicity and because this situation occurs so rarely, many technical report writers would simply add such documents to the list of references unless the report was very formal.

If you do include a bibliography, then use the same style that you used for the references – either Vancouver or Harvard.

3.19 – Appendices

Many technical reports have one or more appendices. Their job is to provide additional or supporting information for those readers who would like to see it, probably the most technically competent. However, remember that your readers should never need anything from an appendix in order to understand the report and accept the conclusions.

Appendices can contain text, diagrams, tables, photographs and so on – anything that can provide supporting information. They could include complex calculations, extensive descriptions of methods, measurement data for graphs displayed in the report itself, and so on. You can think of them as storage boxes for interesting but non-essential information. By using such storage boxes you make the main body shorter and less cluttered, and probably clearer. Of course, every appendix must be referred to in the report; you put it there so there must be a reason.

As previously mentioned, sections in your report are itemised by number (Section 1, etc.) whereas appendices are itemised by letters (Appendix A, etc.). Appendices may need their own figures and tables, and these are referred to as Figure A1, Figure A2, Figure B1, Table A3, etc.

Appendix or Annex

Do you need annexes as well as appendices? Are they different? Many people find this confusing and call everything an appendix, but they are different.

Appendices are written by the same author or team that wrote the report. They belong with it and would not necessarily make much sense as separate documents.

Annexes are independent documents that make perfect sense without the report and have lives separate from the report. They are not part of the report and they are nearly always written by someone else. They are included for the convenience of the readers. For example, an author may add a manufacturer's specification sheet for some component because it may help the readers.

If you use both appendices and annexes, put the appendices first as they belong to the report.

3.20 – Chapter Summary

Format your report using your organisation's template. If there is no template, then use the format suggested above or one of the other formats we discussed earlier, such as the managerial format or the 1:3:25 format. Modify the format as necessary bearing in mind that few reports need every section we have looked at and even fewer need every item we have considered.

Decide early if your introduction will include the methods or if they deserve a separate section. Normally, make the aims a subsection of the introduction with their own subheading.

The results should be purely factual and the analysis (if there is one) must be dispassionate. The discussion should explain what it all means, giving your considered professional opinion or opinions. If there are competing options, discuss their relative merits. Do not brush aside any conflicting results but instead consider them and explain why you have chosen to set some aside, if indeed you have. As non-technical readers may read it, the discussion should be understandable to a wider readership than the results and analysis. They may not want all the gory details but they do want to know what they mean – the implications.

Let the conclusions arise naturally within the discussion if they do, but then present them boldly and concisely in the conclusions. Some people will meet them there for the first time, others will already be familiar with them but want to see them all in one place. The recommendations should also be bold and concise.

If you have references, decide whether to use the Harvard or Vancouver system. In my experience, the Harvard system is used more widely than the Vancouver in technical reports.

4

WRITING THE DRAFT

You have now planned your report in considerable detail and you have your architect's diagram, which you have agreed. It is time to call in the builders, time to start writing, time to turn that plan into a draft report. With a good plan you may manage with a single draft, although it will certainly need some editing and probably some rewriting.

You may have chosen to plan the entire report and then write the entire report, or you may have decided to do it section by section. Either way, your aim now is to convey your message clearly, concisely and precisely, in good English, in such a way that your intended readers can understand it at their first attempt in exactly the way you meant them to. That is not easy to achieve, in fact it is hard, but it is worth attempting. At least get a rough diamond for now. When you start editing you can cut and polish it until it is a sparkling gem.

A long time ago, in 1905, HW Fowler (1858-1933) and his brother FG Fowler (1871-1918) published a book called *The King's English*. Its modern derivative, *Fowler's Dictionary of Modern English Usage*, is still in print and kept up to date. The brothers gave us this advice:

> '*Anyone who wishes to become a good writer should endeavour, before he allows himself to be tempted by the more showy qualities, to be direct, simple, brief, vigorous, and lucid.*'

Well over a century later, that advice is just as valid as it was on the day it was written. Although the English language has changed a lot since then, the basic rules for good writing remain much the same.

Here is a summary of the main pieces of advice you are likely to find useful when writing your reports. We will explore them below, along with some other points. Please note that I claim no originality for this advice. I am merely passing on advice that I have been given and have found useful. It has been repeated many times, by many writers over many years.

The main advice is:

- Put your readers' understanding first.
- Be clear, concise and correct – and Cut! Cut! Cut!
- Prefer short paragraphs to long ones.
- Prefer short sentences to long ones.
- Treat punctuation as a customer service.
- Do not nitpick over minor stylistic issues.
- Prefer short words to long ones.
- Prefer active verbs to passive verbs.
- Prefer strong verbs to weak nouns (nominalisation).
- Use plain English.
- Use lists, illustrations or tables where they work better than prose.
- Write numbers and SI units correctly.

If you can apply these suggestions (they are not rules) then the quality of the writing in your technical reports will improve noticeably.

We will also look at some other points including:

- Monitoring sentence length, structure and punctuation.
- Some misconceptions including whether to start a sentence with *And* or *But*.
- Using emphasis in a report, and why you should avoid the underline button.
- Using parallelism or symmetry in writing.
- What to do with metaphors, similes and clichés.
- How to cut waffle and superfluous words.
- Coping with jargon and acronyms.
- Different types of list, including bullet points.

It is important to stress that while writing and editing are separate stages in the process of producing a technical report, much of what we discuss in this chapter applies to them both. You will not get everything right first time, not even after some rewriting and editing (I certainly do not), but the quality will improve and that should make it easier for your readers to grasp your meaning quickly and accurately. No one gets everything right first time when writing and everyone needs to do some rewriting and editing. You will even find a few mistakes, mostly typing mistakes, in published books written by professional authors and edited by professional editors.

Vladimir Nabokov, whose most famous novel is *Lolita*, is quoted as having said:

> '*I have rewritten, often several times, every word I have ever published. My pencils outlast their erasers.*'

Many other writers have made similar comments. A modern equivalent might be: my delete key is the first to fail.

It is one thing for professional novelists to rewrite and rewrite, but you are writing a technical report not a novel. You have a deadline and you have neither the time nor, probably, the inclination to rewrite everything several times over. That said, you should devote time to improving your draft until you are confident that your intended readers will understand it in the way you intended them to, ideally at their first attempt.

However, before you can improve your draft, you have to write it. Therefore, the so-called writer's block must be banished. Many people put off starting to write, always finding something else to do rather than start typing. It is a common ailment. A famous British journalist, writer and broadcaster called Bernard Levin (1928-2004) once confessed:

> '*I calculate that I have eaten at least 7,000 tons of digestive biscuits in my time solely because the prospect of eating a digestive biscuit seemed to me to be more inviting than sitting down and hitting the keys of the typewriter.*'

Feel welcome to eat a biscuit or cookie or two, but start writing.

Interruptions are another problem for a writer and they can be a bane. They break your concentration and force a reset. As we are not professional writers, there is always a risk that your writing style will change a little after a break. For that reason, try to prevent interruptions. Also, take your breaks at the end of a section or subsection when you have completed one task and before starting another, and where slight changes in style, if any, will be less noticeable.

Two persistent interrupters are the grammar and spelling checkers. Turn them off. For the moment all they do is disturb you. Who can resist checking what is wrong when they produce their familiar blue, green or red lines? Remember to turn them on again when you start editing because they are valuable tools.

So, with or without digestive biscuits, find a quiet place free from interruptions so that you can concentrate for 30 or 40 minutes at a time. That can be difficult to do in an open-plan office so either book yourself into a meeting room or wear ear defenders. Maybe even put a large do-not-disturb sign on your desk. After a joke or two, people will stop commenting on it and leave you in peace. Thirty to forty minutes is about the right time before your concentration starts to wane.

4.1 – Put your readers first

We cannot overstate the importance of putting your readers' understanding first. If your intended readers cannot understand what you have written, what was the point of writing it?

This is about more than simply using words your readers understand, it is also about how you craft paragraphs and sentences. Writing is a craft (some writers refer to themselves as 'wordsmiths') and like any craft it requires hard work and practice to do it well. Learning some of the tricks of the trade is helpful, which is what this chapter is about. As we mentioned earlier, several famous writers have commented that easy reading requires hard writing, and its corollary that easy writing produces hard reading.

Even the way you organise your report, which we have already considered in detail, puts your readers' understanding first. It is not by chance that the executive summary comes first. Everyone will read it and it gives the essential points. Many people will not need to read anything else. It is not by chance that the logical flow of the whole report helps those who will read all of it to anticipate the conclusions before they reach them. And it is not by chance that the detailed, hard-to-understand content is kept apart from the easy-to-understand content.

Whenever you have a choice of how to phrase something, ask yourself which version will be the easiest for your readers to understand. Ask yourself what they need from you at this point. Ask yourself, 'What am I trying to say and how can I make it easy for them to understand?'

4.2 – Be clear, concise and correct

These are the three C's as some people call them, the three things you are aiming for, to be clear, concise and correct (or as some prefer, to be accurate, brief and clear – ABC).

These three words (whichever version you choose) capture the essentials that make a good writing style. The meaning is clear to your intended readers whether they are technical specialists reading everything or non-technical managers reading only the executive summary. The writing is concise with no wasted words, no waffle and no long-winded ramblings. Finally, it is correct – not just the facts and conclusions – but the grammar, spelling and punctuation too.

Clear

How easy is it to understand your report? Read it carefully. It is amazing how many people write a report but never bother to read and check it.

If possible (and it may not be), read it aloud. That sounds strange but it is the best way to alert yourself to passages that are not as clear as they could be or do not flow as well as they should. Perhaps more realistically, read it silently but quickly. If you stumble, then maybe something is not quite right. Look for it. You wrote it; if you cannot read it quickly without stumbling then your readers are probably going to stumble too.

Try to judge if your intended readers will understand the report easily, without hesitation and without misunderstanding. This is hard for writers to do because they know it so well by this stage, but do your best. Ideally, all your intended readers should understand fully the sections that matter to them at their first reading. This is almost impossible to achieve but try to get as close to that ideal as you can. Much later, when you have completed your editing, someone else should review it by reading it carefully, checking for factual accuracy, clarity and ease of understanding.

Many things affect clarity including jargon, acronyms and abbreviations, the average lengths of paragraphs and sentences, the structure of sentences and the choice of words. We will examine all of these in this chapter but will summarise them here.

Is there unnecessary jargon in the widely read parts of the report? At the draft stage, the answer is usually 'yes'. Remember that your readers for these sections include many non-technical people. Although some jargon will be needed, cut unnecessary jargon – and that may include some acronyms and abbreviations. In engineering we talk in jargon and acronyms, as do all professions. That is fine in the main body of the report where technical precision is vital, but how much has crept into the parts that non-technical people read? Check the executive summary, the introduction, discussion, conclusions and recommendations. Jargon can be a stumbling block for non-technical readers and even for engineers from another discipline. Some argue that one acronym in the executive summary is one too many. Rewrite parts of those sections to make them easier for non-technical readers.

Technical report writers need to strike a balance between a precise meaning, using jargon that perhaps few understand, and a less precise meaning using widely understood words. It is a dilemma because we want both precision and understanding, and there may be no alternatives for many technical jargon words. Sometimes we need to remind ourselves that our non-technical readers do not want a precise understanding of the technical findings, but they do want a precise understanding of the consequences. For them, a less precise technical understanding may be welcome. It may be a good idea to start the technical sections of your report with an overview, a jargon-free zone of a few paragraphs for the non-technical readers, before allowing the jargon to build for your technical readers. It all depends on whether you expect any non-technical people to start reading the technical sections.

Keep your paragraphs relatively short, which is what your planning should be pushing you towards anyway. A visual check will show if any are too long. If there are any, split them.

Do your sentences ramble on and on? Check the average sentence length using the Microsoft readability statistics, which we will look at later. Do this section by section because you can accept longer sentences in the heavy technical parts of the report than in the most widely read sections. As a rough guide, aim for an average of 18-22 words per sentence for the whole report, slightly fewer in the popular parts, slightly more in the technical parts. This book has an overall average of about 19 words per sentence. Do your sentences duck and weave before getting to the point? See if you can cut any of the meanderings and straighten the sentences a bit.

Have you used any pompous or fancy words when ordinary words will do the job? There will be some – change them. Use a thesaurus if necessary.

Concise

Draft documents can always be shortened, often by more than 10 %. Many famous writers have said things like 'Cut! Cut! Cut!' and 'If you really like a sentence, cut it' – and one that sounds rather threatening, 'Kill the darlings!'

They are all great professional authors but be careful. The advice is undoubtedly sound, but you are writing a non-fiction technical report and not hoping to win the Nobel Prize for Literature. Take their advice and amend or cut the worst from your report, such as ugly or stumbling phrases, words and phrases that do not earn their keep, all waffle, long-winded phrases, superfluous information, and most adjectives and adverbs. All those should go but do not spend days agonising over doing it.

There are two grammatical techniques that can reduce the length of your report by anything from 5-15 % and we will look at them in detail later. They are, preferring active verbs to passive ones and, where possible, preferring strong verbs to weak nouns (known as nominalisation). Our technical education has rather blinded us to these techniques. Your grammar checker will spot most of them and we will consider them in this chapter.

Correct

This refers to the accuracy of your spelling, punctuation and grammar rather than to the accuracy of your facts and logic, which we will take for granted here.

Make full use of the spelling checker, set to the appropriate style of English of course. Unfortunately, it will question your intelligence at times by asking you if you really want to write whatever perfectly correct word you have written. But tolerate that because you may be surprised at the number of times it points out where you have made a mistake. You can add to the custom dictionary any specialist words you feel you must use – although not adding them can prompt you to search for an alternative.

It is useful to have a good dictionary available as well. I recommend having a dictionary on your desk and a thesaurus easily to hand in the office. Whether these are books or software is a personal choice.

Punctuation can be troublesome. There are many rules but also many variations. Some dictionaries provide a guide to punctuation as an appendix, otherwise a web search will always provide lots of advice, sometimes conflicting. Your word processor's grammar checker will check a lot of punctuation, usually with an American style which can be slightly different from the British style, but you can change some of the settings. The most important thing to remember when puzzling over a punctuation dilemma is to ask yourself whether this or that will help your readers. Regard punctuation as a sort of customer service, it is there to make life easier for your readers, not to help you pass an English language exam.

Most engineers I have known have not been secret grammarians or, if they were, they kept it top secret. However, you cannot write well without understanding the basics of English grammar. After all, grammar is the rule book for the language, a bit like the Highway Code is the rule book for driving a vehicle. When you learned to drive you had to prove you had a reasonable knowledge of the Highway Code. When you have been driving for several years you will have forgotten some of it, relying instead on remembering the basics plus your experience and instinct. Grammar can be a bit like that except we now have grammar checkers to help us. The grammar checkers built into word processors are a lot better than they used to be and it is a good idea to learn how to use them well. We will look at the Microsoft Word grammar checker in the next chapter.

Spelling, punctuation and grammar are the big three on the 'correct' label, but there are a few other things as well. The minutiae of reports should also be correct. Things like the consistent use of heading styles, figure numbers and captions, table numbers and titles, references, page numbers, hyphenated words, etc. All these should be consistent. Consider asking someone else to do a final check; fresh eyes see things that your eyes miss.

4.3 – Paragraphs

As a rule of thumb, short paragraphs are easier to understand than long ones, but it is the average length that really matters. Aim for somewhere around four to eight lines on average for a standard A4 or US Letter page. If your report will be read mainly on screen, keep the paragraphs a little shorter.

Of course, there are examples in literature of long, exquisitely structured paragraphs that are easy to read and understand, but they are probably the result of considerable rewriting and we do not want to go there. Most long paragraphs in technical reports are a struggle to read. If you have a string of long paragraphs, your readers may lose focus and give up.

As a short, digestible part of a longer piece of writing, a paragraph's few sentences focus the reader onto a single part of the subsection's message. They should not contain several parts of the message flung together. The fact that they stand out on the page gives a visual signal to the reader that here is a single bit of the message for you. Together, a series of paragraphs reveal and build the message in sequence.

Imagine trying to read a report where paragraphs were almost the full length of the page. You would probably start to struggle, no matter how good the content. Naturally, lengths do vary so use your common sense when judging how long is too long. Our guide of about four to eight lines per paragraph for an A4 or US Letter page (or, another way of looking at it, say five to nine paragraphs per page) is just that, a guide.

Expect paragraphs to be longer in the main body of the report where your readers will be technical experts, and a little shorter in the widely read sections, especially the executive summary, the discussion, conclusions and recommendations.

This approach gives you enough flexibility to explore complicated issues while encouraging the reader to stay with you. A quick visual inspection is all you need to check the average length.

Sometimes, a paragraph gets too long even though it is all focussed on a single part of the subsection's message. Look for where the focus changes, even slightly. That is where you can split it, producing two or three short paragraphs instead of one long one.

Conversely, a paragraph of only a line or two demands attention and can be good for making the most important statements.

When we looked at planning, we saw how to click and drag bullet points to form embryo paragraphs. Now that you are writing, that detailed planning will pay big dividends although, naturally, you are bound to make some minor changes to the sequencing as you go along. No plan is ever perfect, but you should now prove to yourself that separating the sequencing from the writing saves time and makes the writing easier. Those bullet points will also encourage you to write relatively short sentences.

Remember that the first paragraph of a section or subsection should introduce the main point of the section or subsection (the heading will already have signalled this). It is the topic paragraph, the scene setter or the attention grabber. Arguably, it is the most important paragraph in the section or subsection and it starts with the most important sentence. See it as a service to your readers because it helps them to decide whether to read this part or not.

For most paragraphs, see if you can give the main point in the first sentence or at least get a key word or phrase into that first sentence. Then follow with the detail in a logical sequence. In the final sentence you may be able to summarise or point towards the next paragraph. (Remember the journalist's approach? They write so that you can stop reading at any point in the article but will still have read a coherent story, but you will have missed some of the details.)

However, a big caution; none of these suggestions are rigid rules so do not expect to follow them slavishly for every paragraph. Of all the suggestions here, the closest to being a rule is: short paragraphs are easier to understand than long ones.

4.4 – Sentences

Sentences are the building blocks of writing and there are two things to get right. First, the average length where, as with paragraphs so with sentences, short ones are easier to understand than long ones. Second, the structure where, unlike with paragraphs, the structuring or sequencing takes place while writing not while planning.

Sentence length

Within reason, short sentences are easier to read and understand than long ones, but it is the average length that matters. As a guide, aim for an average length of about 17 to 22 words per sentence. (The average length in this book is about 19 words.) Try to keep to the lower end, or even a bit lower, in the executive summary, discussion, conclusions and recommendations, and allow yourself to venture to the higher end, or even a little bit higher, in the heavy technical parts and any appendices.

Deliberately varying sentence lengths adds some variety to your writing and prevents monotony – but keep to that average of around 17 to 22 words for the entire report. Longer sentences are useful when you need to describe complex points and in the heavy technical parts of your report you may need to go to 35, 40 or even more words occasionally. When you do use long sentences, examine them afterwards to see if you should break them in two. If you have two or three long sentences together, then try to follow them with a short one to relieve the pressure. However, remember that this is mainly about average sentence lengths rather than individual sentence lengths.

Extremely long sentences can choke your reader, an unpleasant thought. I once read a 105-word sentence in a letter written to Japanese business people. Most of the other sentences had a mere 50 to 60 words. Even I struggled to understand the letter and I hate to think what the intended readers thought of it.

While short sentences are easier to read than long ones, they are also easier to write; a bonus for you. A sentence of 40 or 50 words needs a lot more skill with structure and punctuation than one of around 20 words. Do not be afraid of occasionally using very short sentences of two or three words if the message calls for that. However, having too many short sentences can become irritating.

This general advice about sentence length is common. For instance, the UK government asks its employees to limit sentences to a maximum of 25 words, which they can exceed occasionally but only when necessary. The American Office of Personnel Management looks for an average of 15 to 20 words. Many writing guides also suggest an average of 15 to 20 words. Some see an average of 21 words a 'fairly difficult' and 25 words as 'difficult'. Advice for academic writers usually pitches for about 20 to 25 words.

You would be right to think you have an intelligent and literate set of readers, but you would be wrong to think that significantly changes the guidance. Why? First, because technical reports often discuss quite difficult concepts that require a lot of concentration. So, help your readers as much as you can, which short sentences will do. Second, because your readers are busy people who want to get the message quickly and easily without having to re-read sentences to understand them. Also, they may be reading your report in an open-plan office, or on a busy train or in other places where it is hard to concentrate.

Sentence structure

It is not all about average length though. Structure your sentences in a straightforward way. Start many, or even most, of your sentences with the most important point and then elaborate. This will mean writing in a direct style: subject then verb and then the rest of the sentence. 'The cat sat on the mat,' is (normally) preferable to, 'The mat was sat on by the cat.' (One exception could be in a forensic examination of the mat!) This is known as preferring the active voice to the passive voice, but more of that later. Doing this will never win for you any prizes for literature, but it will get your message across clearly to your readers who want you to get to the point quickly.

As already mentioned, using a single sentence as a paragraph can draw attention to a particularly important statement, but do not do it too often.

There will be many times when you want to include asides or extra information in sentences. As a guide, prefer not to put such clauses or phrases into the middle of sentences as that cuts the main message in two and breaks the reader's thought process. Learn to notice this when you do it and question whether you should rearrange things. This is not a wrong thing to do, it is just that asides are often better at the end, or even the beginning, of sentences. Once again, do not apply this guidance rigidly as there are many times when putting the aside in the middle will work fine, and could be the best place for it.

However, consider this sentence.

> *The maintenance of the rollers, which were supplied by several manufacturers and were serviced by several teams from three independent service companies over the past five years, have not been satisfactory.*

At 31 words it is long but not too long. However, the 22-word aside in the middle of an otherwise short sentence of 9 words makes the readers concentrate a bit harder than necessary. It also hides a grammatical mistake. The subject of the basic sentence, the maintenance, is singular whereas the subject of the aside, the rollers, is plural. The writer correctly used a plural verb in the aside but forgot to switch to a singular verb for the last part of the basic sentence.

The basic sentence should have been:

The maintenance... has not been satisfactory.
not
The maintenance... have not been satisfactory.

This is a common mistake when long asides split an otherwise short sentence.

The aside is so long that even the grammar checker misses the mistake. It may be a comforting thought that many readers will also miss it.

The problem is easily resolved by writing two sentences or by putting the aside at the end of the main sentence.

For example:

> *The maintenance of the rollers has not been satisfactory. They were supplied by several manufacturers and were serviced by several teams from three independent service companies over the past five years.*

Shortening long sentences

In your draft report you will probably have some sentences that are far too long. What can you do about them? Here are some ideas.

Split long sentences in two. The split will probably be where you see some sort of connecting word, which connects two or more thoughts. You may not need to start the second sentence with the connecting word. Connecting words include: *and, but, or, nor, so, yet, after, before, since, because, owing to, due to* and so on. You may have to split some long sentences into more than two new sentences but only if it is sensible to do so.

Cut any waffle and unnecessary words; for instance, do you really need all those adjectives?

It is not true that sentences should only contain a single thought. If you stuck to that idea your writing would be staccato and irritating to read. However, do not overload sentences with too many thoughts; each of which will probably be expressed in a single clause. Too many thoughts mean too many clauses which makes the sentence too long. Sometimes using a list might be better. If all else fails, scrap the sentence and ask yourself what you are trying to say, and then rewrite it.

Measuring sentence length

All this talk about average sentence lengths may have you imagining spending hours counting words and doing the arithmetic. That is what used to happen but now your word processor's readability statistics will do it for you in a flash. Simply highlight the section you want to measure and the result will be displayed for you. How to do this is described in the next chapter.

Punctuation

The purpose of punctuation is to help your readers to understand your writing. Punctuation shows how sentences are constructed and how to read them. If you are puzzled over how to use punctuation, and it certainly can be puzzling at times, remember that guiding the reader towards getting the correct meaning is more important than trying to follow half-remembered rules. The grammar checker in MS Word does a good job at flagging issues with punctuation, if at times it can be a bit pedantic and is based on American style which differs at times from British style. Take note of what it says and then make your own decision.

Some uses of punctuation are about preference or style, but others are either right or wrong. All the punctuation marks are described in the style guide at the end of this book where I have been guided by the punctuation appendix in *The Concise Oxford Dictionary, New Hart's Rules: The Oxford Style Guide* and *The Chicago Manual of Style*. Of course, an internet search will answer most of your questions even though some articles might add confusion rather than clarity – especially as there are some differences between British and American usage. If your organisation has a style guide that differs from other advice, including the advice in this book, then follow your style guide unless you are convinced it is wrong, which it may be. In that case, raise your concern internally.

Four misconceptions

There are several minor points of grammar or style that people argue about. Grammar is the set of rules or conventions for a language. Like punctuation, it helps you to make your message clear for your readers. However, English grammar can be messy. Conventions and styles change over time, minor rules become outdated and not everyone is, or wants to be, up to date. Consequently, there are disagreements. For example, there is no universal style guide so publishers have their own style guides which can disagree over minor points.

Here we look at the four issues that I am asked about most frequently on my courses. While they are based on good sense, they are not compulsory rules of modern grammar although some people regard them as such. You are writing a technical report not a fancy modern novel so do not risk alienating any of your readers who might see these as fixed rules that must never be violated.

Starting sentences with *And* or *But*

Grammatically, you can start a sentence with any word you like. It has never been a rule of English grammar that you cannot start sentences with *And*, *But* or indeed any other word. You will find such sentences in good literature and bad, and in old literature and new. The respected *Chicago Manual of Style* says that up to ten per cent of sentences in first-rate writing begin with a conjunction (of course, much of that writing may be fiction).

However, in technical reports there are two reasons not to do so. The first is the already mentioned risk of alienating readers who believe it is a fixed rule, whether they are a client or your manager. If they might think it is wrong, then let it be wrong. Second, it is very unlikely that you will ever need to start a sentence with either of those words. If you do, try deleting it and see if it makes any difference.

Ending sentences with prepositions

Prepositions are short words, or phrases of a couple of words or so, and are usually about place or time and there are well over 100 of them. Here are some examples: *in, on, to, after, before, from, up, down* and so on. At school you may have been taught not to end a sentence with any of them. Prepositional phrases include *owing to, instead of, such as, apart from, in front of, as well as* and many more.

This advice is highly disputed and many good authors ignore it. Consider the sentence, 'Put your shoes on.' That is fine even though it ends with a preposition. If you could never end a sentence with a preposition, that sentence would have to be, 'Put your shoes on your feet.' That is hardly necessary, is it?

Winston Churchill (who was awarded the Nobel Prize in Literature in 1953) is said to have ridiculed and parodied this supposed rule with a comment along the lines of:

'This is the sort of English up with which I will not put.'

He preferred:

'This is the sort of English I will not put up with.'

That, of course, ends with a proposition. No one seems to be certain of his exact phrasing and several versions have been claimed.

How should you decide whether a preposition at the end of one of your sentences is acceptable? Very simply, does it read well? If it does, leave it; if it does not, change it.

Commas before *and*

This is usually about putting a comma before the final item in a list and is covered later when we discuss how to punctuate lists. You can put a comma before the word *and*. It is often unnecessary in British English although it is common practice in American English. Use one if it will help to clarify the meaning for your readers. It is especially helpful when another *and* is nearby. For example: 'We went to two pubs: the Pig and Whistle, and the Bull and Bear.'

Splitting infinitives

The infinitive is the basic version of a verb. In English, it is always two words, the first being *to* which is followed by the rest of the verb as in 'to measure', 'to examine', 'to check', 'to go', etc. A split infinitive is when we put one or more words between the *to* and the second part of the verb, as in 'to diligently measure', 'to minutely examine', 'to carefully check', and the famous one from *Star Trek*, 'to boldly go'.

At school, many of us were taught never to split an infinitive but times change. Often, it is unnecessary to split an infinitive and you may get a smoother sentence if you do not do so. For example: 'to measure diligently', 'to examine minutely', 'to check carefully' and 'to go boldly'. Do not get overly concerned about this. However, there are times when splitting is the clearest way to express what you mean, as in 'to more than double the sampling'. Here the infinitive 'to double' is split by two words. Try writing that without a split infinitive. (Maybe: 'take least twice as many samples'.)

Be happy to split an infinitive if doing so gives the clearest message, but do not let the practice become an excuse for sloppy writing. The Microsoft grammar checker will flag if two words have split an infinitive, but not if there is only one.

Emphasis

There will be many occasions in your reports where you will want to emphasise an important point and there are several ways to do it.

The beginning of both sentences and paragraphs is more prominent than the middle, so whatever you start with will get some emphasis. Repeating information also provides emphasis and you can occasionally close a paragraph or subsection by repeating (but rephrasing) the main point – in other words, the start and end stand out more than the middle.

A single-sentence paragraph demands attention and is a simple way of emphasising a point provided you can write it in a few words, possibly in a single line.

While single-sentence paragraphs attract attention, bullet points scream for it. They are as effective as illustrations and tables at catching the eye. And, speaking of illustrations and tables, colour can add a strong emphasis to graphs and tables.

Also, there are the obvious emphasis techniques of bold, italic and underline. Use these sparingly. Italics are the preferred form for emphasising text in relatively formal writing such as technical reports for clients. Bold is normally second choice although some style guides restrict its use to internal reports, not seeing it as professional enough for external reports. Underlining to emphasise text is out of favour these days and most writers avoid using it. It is seen as old fashioned, being the only one of the three available in the days of typewriters. Also, some believe it makes reading a little harder as the underline obscures the dangling edge of letters like *j* and *y*. However, underlines are used to highlight links to web sites and that can be regarded as their new role.

Inverted commas, or quotation marks, can be used to enclose an unfamiliar word, especially when introducing it to the reader before going on the use it in the text. However, never do this because you cannot be bothered finding the right word, or as an apology for writing a colloquialism as in 'touched base with…'

Finally, two more things not to do. First, do not use ALL CAPITAL LETTERS for emphasis except for the title or possibly for the highest-level section headings. Second, do not change the font size or font style to emphasise text.

Above all, use emphasis sparingly as it loses its impact if used too often.

Parallelism or symmetry

Parallelism sounds classy but all it means is to express similar ideas or thoughts in the same way, in other words to match the arrangement of specific words, phrases or sentences. It is a simple idea that you may find you do automatically because your mind has tuned into an arrangement as you write. For the reader, it helps sentences to flow and strengthens similarity. Lack of symmetry can produce phrases that clash and make the writing harder to understand.

For example:

- Giving a presentation is harder than to write a report – does not flow.
- Giving a presentation is harder than writing a report – does flow.
- It is harder to give a presentation that to write a report – does flow.

In these examples, the parallelism or symmetry comes from using the same type of verb to describe both presenting and writing: In 'giving and writing' and 'to give and to write' the verbs match. Using different parts of the verb leads to a clash, as in 'giving' and 'to write'.

Parallelism has a wider use than in sentences. For example, if you can start each item in bullet point lists with the same type of word, such as starting all with the infinitive of a verb, a participle of a verb or a noun.

Metaphors, similes and clichés

A metaphor is a word or phrase that is not literally true but compares something to something else that has similar characteristics such as 'These regulations are a jungle.' A simile compares one thing to another by using the words *as, like, than, resembles* and so on. 'The liquid was as clear as crystal,' is an example. Clichés are clever expressions that have become boring through overuse and it is the boring aspect that is the problem. At one time, they may have been vivid similes or metaphors. They were 'in the limelight' but now they are 'as dull as ditch water' so 'avoid them like the plague'.

A technical report is a serious piece of writing with a sharp focus and it is tempting to say you should avoid metaphors and similes and stick to the plain facts and the literal truth. However, it is not as simple as that.

Metaphors and similes are used in everyday language and can help to clarify ideas. Writers use them in reports, perhaps without noticing. If a good metaphor or simile helps to make the message clearer then use it, but avoid common metaphors and similes as they draw attention to themselves rather than clarify the meaning. Anything that distracts the reader from the meaning to the writing style is bad news. We want the focus to be on the message, not the medium (and that is dangerously close to being a cliché). Therein lies the problem. As analogies, good metaphors and similes can help to make the meaning clear but often they are not as good as the writer thinks. Be careful when using them.

To sum up:

- Aim for an average sentence length of about 17 to 22 words over the entire report.
- Aim for a lower average length of about 15 to 20 words in the executive summary, discussion, conclusions and recommendations.
- Aim a little higher, about 20 to 25 words, in the heavy technical parts of your report.
- Use longer sentences, if necessary, when explaining complex points.
- Structure your sentences in a straightforward way, often starting with the subject followed by the verb and then the object.
- Take care when putting asides in the middle of sentences.
- Use emphasis sparingly and prefer italics to bold; avoid underline except for hyperlinks.
- All of these are guidelines, not rules.

4.5 – Words

Writers are often advised to prefer short words to long ones. While this is reasonable advice and a snappy slogan, it is not the heart of the matter. The core advice is to avoid fancy, clever-sounding words purely for the sake of sounding clever. The advice should steer you away from words that may be clever but obscure towards plain English words – provided they do the job well. Think of plain English words as being words that the intended readers understand.

Searching for a better word to express your meaning is good news provided it is a search for a word that makes it easier for your intended readers to understand your meaning correctly. When it becomes a search for a fancy word to show off your vocabulary, then it is bad news for the reader. It is also bad for you as research suggests that readers can see it as pompous and it may start to dent their confidence in you.

The reader's understanding is what matters. Frequently, that will mean using long words that catch the meaning precisely, and some technical words are long. If a shorter word will do the job just as well then prefer it, but often there will not be a direct equivalent. This hardly matters in the main body of your report as your readers will be specialists.

However, in the widely read sections it could be a problem and you may have to use several non-technical words to solve it. At least explain the meaning of the technical jargon to the non-specialists; they may be the ones making the big decisions. It is pointless using words you suspect they will not understand and then shrugging your shoulders. Searching for the meaning of the jargon word on the internet can give you examples of how other people have described it.

Easy reading does come from hard writing and you may need to work hard to explain difficult concepts using plain English. At least ask yourself, 'What am I trying to say?' Here are two ideas that might help you (although clear thinking is the best trick).

The first idea is to choose one, and only one, term for any technical object or process in your report. Choose the specific word that is right. If you must use it in one of the widely read sections, then carefully explain what it means. To do this, use a good dictionary and thesaurus (all writers use dictionaries) and check the definition on several websites. Between them you should get the basis of an explanation you can use. Picking one word for an object or idea and sticking with it avoids confusion. While the English language has a rich vocabulary, a technical report is not the place to explore its boundaries.

The second idea comes from a comment attributed to several famous people including Albert Einstein, '*If you can't explain it to a six year old, you don't understand it yourself.*' How could you explain this to a child? Apart from anything else, it is a good mental exercise to take something that is complex and explain it in a simple way.

There is another idea, but it comes with a big warning because sometimes it can be a thoroughly bad idea. You may get bored with using a single word for something, even though it is the right word. Sometimes you feel your readers would like a change too. Normally, resist the temptation. However, if – and it is a big 'if' – if you have already clearly established the meaning for your readers, you have fixed it firmly in their brains, then you may be able to use an alternative word. There may even be a hierarchy of terms to choose from, moving from specific to general, such as Apple iPhone, iPhone, smartphone and mobile. Only use this hierarchy idea if your readers are familiar with all the terms and will not misunderstand. Each successive term is less precise than its predecessor so be very careful about how far you go.

Remember that this comes with a big warning as disaster can await if you go too far along the chain, that is when the reader starts to wonder if you are still talking about the same thing.

Spelling

The development of the English language over the centuries has produced some very strange spellings. Use your spellchecker and make sure it is set to the appropriate version of English but keep a dictionary handy as well. When readers notice a spelling mistake, they may assume it was a typing error rather than ignorance – and there are sure to be some of those in a long report. As we said earlier, even books contain typing errors and they have been written, edited and proofread by professionals.

When your readers notice many spelling mistakes, they may start to think it is because of laziness. Once that seed is sown it can grow quickly to become a weed that strangles their confidence in, not just your writing, but your entire report. Check your spelling. Especially check the spelling of names, organisations, equipment, devices, components, chemicals, processes and so on. Triple check any words you add to your custom dictionary.

4.6 – Active and Passive Verbs

One universal piece of advice given to technical writers is to prefer active verbs or sentences to passive ones. You may have missed this bit of grammar at school, but it is quite simple. There are two ways to write most sentences.

For example:

- Active (6 words): Table 3 shows the pipeline properties.
- Passive (8 words): The pipeline properties are shown in Table 3.

These are two simple sentences, but they illustrate the difference well. You will have far more complicated ones than these in your reports. The first sentence is known as an active sentence and is one where the 'doer', the active one (Table 3) starts the sentence. The second is a passive sentence and is where the object to which something is done (The pipeline) starts the sentence.

Active sentences are usually shorter, more direct and slightly easier to understand than their passive equivalents, although often only by a fraction. That claim is easier to see in long sentences than in short ones, but it can have a large impact if the passive style is predominant in a report – which it often is in technical reports.

Here are some more examples.

- Active (14 words): The research confirmed that sugar coating the tablet does not affect the medicinal properties.
- Passive (17 words): It was confirmed by the research that sugar coating the tablet does not affect the medicinal properties.

- Active (10 words): A separate geological report provides information on the seabed features.
- Passive (12 words): Information on the seabed features is provided in a separate geological report.

- Active (11 words): Different manufacturers installed them and they are over 20 years old.
- Passive (13 words): They were installed by different manufacturers and they are over 20 years old.

Please do not think that the difference is always about two words, it can be far more than that but often it is only about two words. Even at that rate, actives are preferable because we want brevity and it is a bonus if they are more direct as well. Depending on how many passives you habitually use, changing some passives to actives where sensible could easily cut 5 % of words from a report, probably more.

The following example shows the big impact that changing passives into actives can have:

- Passive (21 words): A recommendation was made by the team that all ten-year-old attenuators be replaced by the end of the month.

There are two passives in this sentence: 'was made' and 'be replaced'.
- Active (15 words): The team recommended replacing all ten-year-old attenuators by the end of the month.

Most engineers overuse the passive because it is a spin off from our scientific education. If I could remember the first report I ever wrote, it was probably as part of some chemistry homework written when I was about 11 years old. It probably started something like this: 'We filled a beaker with water and boiled it over a Bunsen burner. We added potassium permanganate crystals and...'. After marking it would have looked like this: 'A beaker was filled with water which was boiled over a Bunsen burner. Potassium permanganate crystals were added...'.

Oh dear! All my nice active sentences (which I had never heard of) had been changed to passive ones (which I had also never heard of) because in science writing we avoid using the pronouns *I*, *you*, *he*, *she*, *we* and *you*. (We do use *it* and *they*.) You and I learned this lesson well. Ever since, in technical reports, we still avoid pronouns although some organisations allow them in internal reports. Worse, we have learned to love passives and use them here, there and everywhere when we do not need them.

Non-technical, professional writers use fewer passive sentences than we do. Some aim for around 10 % to 20 % of sentences being passive. That is partly because of their subject matter but it has a lot to do with their professionalism. I have seen many technical reports where around 80 % or 90 % of sentences have been passive, which is far too many.

There is no rule as to what percentage of your sentences should be active or passive because technical reports vary so much. However, try to aim for passive sentences being between something like 25 % and 45 % depending on your report. In some technical subjects you may need more than that. There will be many instances where you need passives and you should use them, but try to avoid them when they are not needed especially in the widely read sections of your report like the executive summary, discussion, conclusions and recommendations. Roughly 12 % of the sentences in this book are passive.

Passives, though, do have some advantages. They can place the emphasis on what was done rather than who did it, which for much of the core of a technical report is exactly what you want. Therefore, expect to use more passives in the heavy technical parts of your reports than in the widely read parts.

Another advantage is that you do not need to name the 'doer' as in 'several options were considered', 'five tests were conducted', 'two circuit breakers were fitted' and 'the cause has not been identified'. You will use a lot of expressions like those and they are great. Not naming the 'doer' is helpful if you are reporting mistakes as in, 'mistakes were made' rather than 'X made mistakes'. It is never a case of passives are bad and actives are good, simply that actives are shorter and a bit more direct, so prefer them where you can.

Spotting and converting passive sentences

If you are using Microsoft Word then the built-in grammar checker will highlight passive verbs for you. Words will be underlined and there will be a message like 'Passive voice' or 'Consider using active voice' or 'Saying who or what did the action would be clearer.' It all depends on which version of Word you have. You may need to right-click the highlighted words and select 'Grammar' in the pop-up box to see the message.

If this does not happen, go into the settings for the grammar checker (Tools/Spelling and Grammar/Options). At Microsoft, they seem to have a habit of frequently changing what happens next. Select the style that checks as many things as possible – look in the settings to see if they include passive voice. Some older versions of Word offer a 'Technical' style, which sounds appealing. Ignore it because it does not check for passives. In that version of Word, chose 'Formal' as that checks everything.

Once you have identified a passive then you have to decide what, if anything, to do about it. The grammar checker will often offer you an alternative but be careful, it does not always do a good job and it is better to take charge of this conversion yourself.

If you imagine yourself holding a pen or pencil in your fingers and flicking it end to end by 180 degrees, that is what you are about to do with a passive sentence. Note that you may only need to flick one part of the sentence, not all of it. Many sentences have two or three clauses (think of a clause as a phrase with a verb in it) so there could be two or three verbs in one sentence. Any of them could be a passive looking for a more active life. Look back at the samples given above, we simply changed them round – the beginning became the end, and the end became the beginning. Simple!

Simple? Well, not always. Consider this example where the meaning is the same but the emphasis is changed.

- Passive – Human genome research is being conducted at Barnes College.
- Active – Barnes College is conducting human genome research.

Notice that the passive sentence focuses on the research not the college, whereas the active equivalent puts the focus on the college not the research.

Changing the emphasis is a common problem when converting from passives to actives so take care that you do not shift the focus away from where you want it to be. Always check that the meaning is what you wanted and that the sentence is better, which probably means shorter and a bit clearer.

One other point to consider is that there is usually a small shift in the formality of the writing when passives are changed to actives. A report that predominantly uses passive sentences will feel a bit stiff and more formal than one that predominantly uses active sentences. It is a small but real difference, which you can turn to your advantage.

We have already noted that you will use more passives in the core of a technical report where all the heavy technical writing is, and that little extra formality can add a touch more gravity or authority to the writing. This is where you might use up to around 45 % or so passive sentences, although that figure is a bit arbitrary. In essence then, use fewer passives in the widely read parts but more in the heavy technical core of the report, and generally use fewer passives than you have been used to. You may like to run the readability statistics (see Chapter 5) on some of your older reports to get an idea of how many passives you habitually use.

Our legacy of banning most pronouns leads to more passive verbs and a small but noticeable increase in formality, which may be why some organisations allow all pronouns in internal reports. Allowing pronouns increases the percentage of actives, which leads to more personal and informal reports. That may be the wrong approach where the wish is for a strong and authoritative feel such as with technical reports from engineering consultancies for external clients. On the other hand, a company that wants to portray a new, fresh-thinking, tradition-breaking image may choose the opposite approach with lots of pronouns and active verbs to get reports with a more informal and friendly feel.

In summary: the advantages of actives are that they are usually shorter than passives and they make your writing tighter and your report shorter. They can also make your report a bit less formal without becoming informal. The advantages of passives are that they can focus on what was done rather than who or what did it; they can omit naming the 'doer', which reduces repetition; and they are a little more formal, which can add a slightly stronger feeling of authority. However, they do use more words. Striking a balance is the key. Engineers have a built-in tendency to overuse passive verbs so convert them to actives when it will help the readers.

It is not a case of actives are good and passives are bad, and never has been, just that we tend to use too many of them. Aim to reduce your use of passives to around 25 % to 45 % overall, but the actual figure will depend on where you are in the report. While those percentages are a bit arbitrary, at least learn to use fewer than you have been used to using.

4.7 – Nominalisation

This weird-sounding name refers to another common technique that we technical report writers unwittingly use far too often. This is the use of nouns (things or processes) instead of verbs. For instance, instead of writing 'the use of' in that sentence (*use* in this sense is a noun) I could have written the verb *to use*: 'This is to use nouns (things or processes) instead of verbs.' Even better, 'This uses nouns (things or processes) instead of verbs.'

Here are some more examples.

- The attempt was a failure because... Noun = failure.
- The attempt failed because... Verb = failed.

- The company will continue with the development of... Noun = development.
- The company will continue to develop... Verb = to develop.

- The vaccination of 80 % of children prevented further outbreaks. Noun = vaccination.
- Vaccinating 80 % of children prevented further outbreaks. Verb = vaccinating.

As you can see, the argument is the same as for preferring actives to passives. The nouns need extra words to escort them while the verbs sail along happily by themselves. Therefore, preferring verbs to nouns, when verbs will do the job, produces shorter sentences. Also, the verbs produce stronger and more direct sentences, which is usually preferable. Remember that brevity is one of our goals. Again, as with preferring actives, preferring verbs tends to give us slightly less stiff and formal reports – but not informal.

Yet again, do not take this too far. It is not a case of nouns are bad and verbs are good. There will be many occasions where you want to, or need to, use a noun. One example is when discussing a process or technique. Here you will want your focus to be on the process itself rather than using it. For instance, if you were reporting on an 'analysis' of something then you would use that word more frequently than, but not to the exclusion of, its verb equivalents, *to analyse, analysing, analysed* and so on.

Once again, our science education has led us to overusing nouns where verbs would be better. The simple fact is that, as engineers, we have acquired the habit of using nouns without even thinking about using verbs. Let us break that habit while noting that we do need quite a lot of the nouns in the heavy technical parts of the reports. Break the habit and use verbs when you can, even in the technical parts but especially in the widely read parts. Also, yes, once again, take care not to change the meaning or the emphasis of the sentence. If that happens, use the original.

Nominalisation is the grammatical term for nouns made from verbs (they are also made from other types of words) but they are often nicknamed 'hidden' or 'smothered' verbs and even 'zombie nouns' (they suck the blood out of nice verbs). Many of them occur in passive sentences and by converting a passive into an active you may also convert a noun into a verb.

Simply by preferring actives to passives and preferring verbs to nouns (without overdoing either) you can reduce the number of words in your report by something like a 5 % to 10 % while making it clearer and easier to read. Reading a couple of pages where every sentence is passive, and every verb has been smothered into a noun, is hard work compared to the opposite. Aim for the middle ground.

In summary: verbs shorten sentences and give a slightly more direct style than nouns while reducing any stiff formality, so making the report clearer and easier to understand. Nouns can emphasise a concept or process (such as the analysis) rather than the action (analysing) and give a more formal and arguably more authoritative style. Avoid using either to extremes.

Spotting and converting nominalisations

Recent versions of the Microsoft Grammar Checker will find nominalisations for you but look in the settings to be sure they are included. Older versions of Word do not check for them, but there is an easy way to spot most of them.

Many of these smothered verbs (the nouns) have some common

suffixes (word endings) that you can find by using the Edit/Find function. Probably the most common suffixes in technical reports in general are *-ion, -tion, -ment, -ance, -ence* and *-ity*.

Some examples:

ion/tion: recommendation, evaluation, consideration
ment: agreement, arrangement, assessment
ance/ence: performance, inference, resistance
ity: receptivity, transferability, maintainability

Not all smothered verbs end with these suffixes. No doubt there will be other suffixes that are frequent in your subject area and you will get to know them.

Another clue is that the words *the*, *a*, or *an* often precede these smothered verbs and *of* often follows them, as in 'the provision of', 'an analysis of', etc. A combination of these indicators is a strong sign that a verb could replace the noun.

If you use an old version of Word, then use the Edit/Find function to search for words ending with those suffixes that are followed by *of*. In other words, search for *tion of*, *ment of*, etc. with a space before and after the word *of* (so it only looks for *of* and not *often* and so on). You will not find every smothered verb (nor do you need to) but you will find a lot of them, more than enough to polish your style and shorten your report.

Nominalisations can occur anywhere in your report but, as we noted, there will probably be a lot of them in the technical parts – the methods, results and analysis sections. Most engineers will tolerate them there and you will need to keep quite a lot of them, but removing those that are not needed will improve the writing.

If time is short, target the most widely read sections: the executive summary, introduction, discussion, conclusions and recommendations. Also, you could limit your search to the *tion* and *ment* endings and leave it at that.

Not every word that has one of these suffixes is a smothered verb. Using the Edit/Find function will find many other words that are not a problem, so ignore them. If they are preceded by *the*, *a*, or *an* – or followed by *of*, then they probably are smothered verbs, many of which are waiting to be culled.

4.8 – Plain English

Plain English is not simple English and its meaning is imprecise. It could be described as using words and phrases your readers understand easily, or as writing that is clear, concise and unambiguous. For many people, it includes the principles we have already discussed, such as keeping the average sentence length relatively low, preferring active verbs to passive ones, and preferring strong verbs to weak nouns (nominalisation). It also avoids pomposity and verbosity. (Is that last sentence written in plain English? Should it have been something like, it also avoids fancy words and using too many of them?) Overall, plain English provides guidance (not rules) which, if followed, will make it easier for a motivated reader to understand the writing.

The English language is rich in its choice of words. There are usually alternatives, often several and often simpler, but it is not a case of choosing the simplest or shortest word. Sometimes long words have more precise meanings than their shorter equivalents. We need accuracy and precision in technical reports so do not be afraid of using and reusing the same word when it is the right one. As already mentioned, it is normally better to stick to one word throughout the report for specific objects and processes than to try to expand your readers' vocabulary by finding alternatives. If you were writing a romantic novel then you could risk flamboyant language, but in technical reports plain English is seen as a virtue – even if we cannot define it precisely.

As an example, in a report about a landslip it said: 'There was heavy precipitation in the preceding days.' There is nothing wrong with that except that, 'There was 25 cm of rain on each of the previous two days,' would have been more precise and more accurate, and in technical reports we want precision and accuracy.

Waffle

The following list gives some examples of common waffly phrases that creep into reports where something simpler would have been better. I am not suggesting that it is a disaster if you use such phrases, just that you probably do not need them and should try to weed them out. Prefer the plain but use the fancier versions if you really need them.

Here are a few examples:

- a majority of = most
- despite the fact that = despite
- due to the fact that = due to
- during the course of = during
- either of the two = either
- if this is not the case = if not
- if this is the case = if so
- in the eventuality of = if

- in conjunction with = with
- in the light of the fact that = because.

Superfluous words

Adjectives and adverbs are rich sources of superfluous words although there are many times when you do need them. When you notice an adjective or adverb just ask yourself if it is adding any useful meaning.

Many adverbs end in -*ly*: 'usually', 'frequently', 'rarely' and so on. They modify verbs or adjectives and there are normally some in technical reports. Often, they qualify the meaning without being specific. Perhaps you should scrap the adverb and be more specific. How usual, how frequent and how rare?

Adjectives modify nouns and you are more likely to use them than adverbs in technical reports. Keep the ones that add meaning but delete those that do not. Sometimes you will need a collection of them to achieve precision as in:

- the differential thermal expansion coefficient
- the A-300 airframe weight calculation error percentage
- the polyurethane interlayer bonds.

Here are some examples of superfluous words taken from reports of various kinds:

- a serious crisis = crisis
- as to whether = whether
- real danger = danger
- active consideration = consideration
- few in number = few
- entirely destroyed = destroyed
- assemble together = assemble
- brief in duration = brief
- definite decision = decision
- main essentials = essentials
- part and parcel of = part of
- plan in advance = plan
- prerequisite condition = prerequisite
- shuttle back and forth = shuttle
- surrounded on all sides = surrounded
- irreducible minimum = minimum.

Such expressions may be common in speech but cut them from your written reports.

Jargon and acronyms

Jargon, which includes acronyms, is technical shorthand used by one professional when communicating with another in the same field. Some jargon is recognised by specialists in other fields and some passes into the public domain, even if it is not always understood correctly (consider the medical term *chronic* which means long lasting, not bad). Which jargon you may use in which section of a technical report is a question of judgement and should always be decided by asking yourself whether the expected readers will understand it.

Although this has been covered before, it is worth repeating. Use as little jargon as you can in the widely read sections including the executive summary, introduction, discussion, conclusions and recommendations. Some organisations ban acronyms in the executive summary and that is a good position to take. Some also ban them in the conclusions and recommendations, and you should consider doing that if your readers may not know them. You can probably use a little more jargon in the discussion section but remember that some non-specialists will read this section so only use jargon if necessary and be ready to explain it. In the methods, results, analysis and appendices you should be able to use the normal jargon of your specialism. However, subordinate all these suggestions to the question: will my intended readers understand it?

The usual guidelines apply when writing acronyms, that is to write the meaning out in full on first use and put the acronym immediately behind in round brackets. For example: 'The ultra-violet (UV) radiation was intense at this time.' However, it may not be as simple as that. If you introduce an acronym in the methods section and then use it in the conclusions, you may have new readers meeting it there for the first time. As insurance, at least consider redefining any acronyms that you use in the conclusions and recommendations. Even some specialists may not be as familiar with them as you think.

4.9 – Lists

There are three main ways to present lists in technical reports: as a running list within the text, like this one; as a vertical (or 'displayed') list indented and highlighted with numbers, letters or bullet points; and as a left-aligned vertical or displayed list without any indents or highlighting. While lists are useful, try not to overuse them.

In-text lists

A list that occurs within a sentence is still part of the normal sentence and follows the usual rules of sentences. They are called in-text, run-in or running lists. When using them you need to decide five things: whether to introduce the list with a colon, whether to separate items with a comma or a semicolon, whether to use a comma with the *and* before the final item, whether to use *the*, *a* or *an* to introduce each item in the list, and whether to identify each item with a number or letter.

First, should you use a colon? Use a colon when the opening words are the sentence's main clause, which is why I used a colon in the previous paragraph. 'When using them you need to decide five things' is the sentence's main clause (a phrase with a verb) and would work as a stand-alone sentence, hence the colon.

Do not use a colon when the opening words run into the first item in the list. For example, 'The author has to decide whether to introduce the list with a colon, whether to separate items with a comma or a semicolon, whether to use a comma with the *and* before the final item, whether to use *the*, *a* or *an* to introduce each item in the list, and whether to identify each item with a number or letter.'

Second, if the items in the list are relatively short then separate them with commas. As they become long or complex you may upgrade to semicolons. This has the advantage of reserving commas for use within individual items. For example, 'There were four safety issues: damaged bolts, some of which were rusted through; screw heads painted over, particularly countersunk screws; rotting timbers, including some on the standing platform; and slippery surfaces, the worst being on ladders.'

Third, using a comma before the word *and* that announces the final item in a list is a question of style, not grammar, and it differs between British and American English. In British English we would normally leave out the comma as in, 'We bought eggs, milk, butter and cheese.' American English normally includes it as in, 'We bought eggs, milk, butter, and cheese.'

Even in British English there are many times when the extra comma helps to make the meaning clearer for your readers, which is why I am happy to use it in my writing. It is particularly helpful when lists contain lengthy items. As I have mentioned earlier, punctuation is about customer service.

Consider the following example: 'In order of preference, the potential suppliers are Greys, Browns, Reds and Blues, and Greens.' The extra comma before 'and Greens' makes it clear that 'Reds and Blues' is a single company. Without that extra comma readers may think there were five companies involved – or three if they thought 'Reds and Blues and Greens' was a single company, rather like a group of solicitors. If that list had not been in order of preference, then putting the double name at the beginning would also have solved the problem: 'The potential suppliers are, Reds and Blues, Greys...'

Consider another example: 'Other possible causes of the failure are weld or material defect, corrosion, settlement, and fire due to ignition of the tank vent by lighting.' Once again, the comma before the 'and' clarifies the meaning – the settlement and the fire risk are separate issues.

In short, use a comma before *and* if it will help your readers to understand what you are saying. (Commas before *and* in lists of three or more items are called Oxford commas in Britain and serial commas in America.) If your organisation does not give guidance about this, then choose the British or American style as appropriate but be willing to use a comma where it will help your readers.

Fourth, should you use *the*, *a* or *an* to introduce the items in the list? The answer will depend on the context and your emphasis. Using one of them at the end of the introductory statement works well and saves repetition at the start of each item, but putting one of them before each item emphasises their individuality. In some lists, you may not need them at all.

The fifth and final point is about giving each item a number or a letter. Doing so will make it easier to refer to them individually in the text but may make the list fussier. As always, ask yourself which is best for the readers. Strike a balance and do it if you are going to refer to the items several times in the text, in which case it will help your readers. Notice also that using a number or letter strengthens the fact that there is a sequence or order to the list. It implies that the list has a hierarchy or priority, which many readers will infer whether you want them to or not.

Vertical or displayed lists

Bullet points are the most common of the vertical or displayed lists. The others are lists that use numbers or letters instead of bullets, and simple lists that are aligned left with the margin without numbers, letters or bullets.

All have a big advantage over in-text lists, which is that they catch the eye of the reader. Like an illustration or a table, they cry out, 'We are important.' Also, they break up the text, just as illustrations and tables do, and this adds variety and helps the report to look better. Do not overuse them though as they can then lose their impact.

You will probably see bullets as the obvious choice, but they are not always the best choice. If you use numbers (or letters for that matter) many, maybe most, readers will infer a hierarchy which they would not do with bullet points. So, prefer numbers if you want to imply some sort of priority or sequence, otherwise use bullets. Both bullet points and numbered lists need an introductory phrase or sentence.

Simple lists, the third type, are used less often than the other two in technical reports. A timetable is one example. They also need an introductory statement. Unlike bullet points and numbered lists, they are fully aligned left with no highlighting.

There is no universal agreement about how to punctuate vertical lists but there are two common styles, one for lists of short phrases and the other for lists of sentences. Both have variations. You may use both styles in a single report although it may be better to stick to one in short reports if you can. The short phrases version will cope with most situations. Never use both styles in a single list.

Bullets using phrases

Here is an example of a bullet point list of short phrases.

The most urgent proposals concern:
- sterilising equipment
- verifying measurements
- calibrating test equipment.

Note that the three verbs are the same style – ending with *ing* – and the writer has avoided using the nominalised version – sterilisation, etc. A consistent style of phrasing for the list items is important and probably as important as how you phrase them. It is an example of the symmetry or parallelism that we mentioned earlier.

In short bullets, the opening phrase plus the entire set of bullet points (or numbered points) form a single sentence, beginning with a capital letter and ending with a full stop. The introductory phrase ends with a colon – never a colon and a dash. The introductory phrase must perform naturally as the start for each item, in other words you could mentally put it in front of each item in the list. Essentially, short bullet points are an in-text list spread over several lines instead of lost within a paragraph.

Each item can end with a semicolon, a comma or nothing. At one time, semicolons were almost universal and were carried over from in-text lists. Many organisations replaced the semicolons with commas several years ago, and many have now dropped the commas as any punctuation mark at the end hardly adds any value.

The first word of each item begins with a lower-case letter, after all it is in mid-sentence. However, many people like to start with a capital letter for appearance sake. Recent versions of Microsoft Word automatically use a capital letter, which some find irritating and dictatorial.

The final question is about whether to end the list with a full stop. As the list is a single sentence, a full stop brings the sentence to a natural end. However, some organisations now omit it believing that it looks odd when no other item has a punctuation mark at the end. If you use commas or semicolons after the other list items, then you must end with a full stop.

Otherwise, it is your choice. Although I have no experience of this, I am told that omitting the full stop at the end of a list can confuse screen reading software causing it to read the next paragraph as a continuation of the final bullet point. That is why I have ended all my lists with a full stop.

Artistic presentation is important because of the strong visual impact of vertical lists, which after all is why you are using them. Not using a full stop after the final item in a list of short bullet points might be more visually pleasing than using one. Sometimes in bullet point lists the grammar checker can flag an error message if it thinks you have not ended a sentence with a full stop or have not started one with a capital letter.

With all these things, make your choice and apply it consistently.

Bullets using sentences

Here is an example of bullet points using full sentences.

These are the most urgent proposals:
- Sterilise all equipment immediately before use.
- Verify all measurements.
- Calibrate test equipment before and after measurements.

Every item here is a sentence and they all start with a capital letter and the same style of verb, and end with a full stop. The introduction may be a phrase or a sentence preferably ending with a colon, although some organisations us a full stop if it is a sentence.

Consistency

Consistency is important with all lists. All aspects should be as consistent as you can make them across all the items in the list and between lists of the same style. These include starting with either a noun or a verb, using the same tense if starting with a verb, choosing actives or passives, choosing infinitives or present or past participles, using positive or negative statements, using a capital or lower case first letter, using commas or not, etc. Note that with both sets of bullet points shown above the writing style for each item in the same list is the same. That is good practice.

4.10 – Illustrations

Whether a picture is worth a thousand words is debatable, but illustrations are vital to most technical reports as they help readers to understand the message in ways that words cannot do. Their role is to illustrate your message, not to decorate it. That is, they are there to inform, not to make the report look attractive. For our purposes, illustrations include graphs, charts, diagrams, drawings, photographs and so on. These are excellent at visualising trends, equipment, assemblies, components, circuitry, layouts and so on, and there are many styles such as flow diagrams, line drawings, cut away diagrams and exploded views.

Some illustrations provide the main act, providing the message in a way that you cannot do with text. Others are there as supporting cast, helping readers to understand the message in the text. Each illustration must play its part in the story, just as each paragraph does, so illustrations illustrate what the text is saying while the text describes what the illustration is showing. They are mutually supportive. Just as the story is incomplete if one paragraph is missing, so it is incomplete if one illustration is missing.

Plan your illustrations when you plan the report or section. Decide which illustrations justify inclusion and whether they will lead or support the text, and where they will go. You are probably doing this already to some extent without being aware of it, but a little awareness should help you to do it even better.

Place them as near as you can to the text that discusses them as that helps the readers to switch their attention between text and illustration. Also, leave one or two blank lines above and below them. Putting large illustrations close to the text may be impossible, which then gives you a dilemma. In some reports, large illustrations are placed either at the end of the section or at the end of the report. If you decide to do that, then put all the illustrations there rather than mix and match with some in the text and some at the end. That is confusing.

Use new illustrations as far as possible rather than stock illustrations. Keep them simple by asking yourself what point you want them to tell the readers and then make sure they emphasise that point. For instance, if you have several curves on a graph then distinguish them by using coloured, continuous, dashed or dotted lines so that you can refer to them individually in the text. Label them and, of course, clearly label axes, scales and units. Note that horizontal labels are easier to read than vertical ones.

Every figure or illustration must have a figure number and a caption. The numbering style must match the system you are using for the tables, although tables and illustrations have their own sequence of numbers. Either number figures sequentially through the whole report (for instance, Figure 8 could be the second figure in section 3) or use a section number followed by the figure number within the section (in which case our Figure 8 might now be Figure 3.2). Figures in appendices have their own set of numbers preceded by the letter of the appendix: Figure A2, Figure B5, etc.

If your organisation has not provided a style guide, then decide how to punctuate the caption. Preferably, write the word *Figure* in full, starting with a capital letter. Only use *Fig.* (with a full stop) if your style guide tells you to. Write the number as a numeral, not a word, and prefer to put a full stop after the figure number.

For example:

Figure 3. X-ray diffraction pattern of sample 23.

Although these are minor details, you should choose a style and apply it consistently.

Captions are names or titles for illustrations so prefer to use short phrases (i.e. no verbs) for the captions rather than full sentences, as in the Figure 3 example above. They can be as long as necessary but think short rather than long. The words should simply name the figure. If you want to add more information, then use a sub-caption (with verbs) to highlight the main point you want readers to learn from the figure. Write sub-captions in italics or put them in round brackets so that they do not compete with the main caption.

For example:

Figure 3. X-ray diffraction pattern of sample 23. *The fuzzy circles prove that this sample is amorphous, not polycrystalline.*

By convention, place figure numbers and captions below the figures. Align them either left or centre and be consistent.

Finally, you have more freedom to use abbreviations in figures than in text because space is limited, but do not use them unnecessarily. This includes using the ampersand (&) which you would not use in the text.

4.11 – Tables

Tables are excellent at showing detailed data in an organised way, whereas graphs or charts are good at showing trends. The details that a table displays vividly would be lost in the text or on a graph. While tables can display a lot of complex data clearly, keep them as simple as possible consistent with what your readers need. Additional data could go in an appendix.

Help the tables to stand out from the text by leaving one or two blank lines above and below the table.

Arrange the columns and rows logically in a way that the reader can understand easily. Centre align the column headings but left align the row headings. Try using bold for the headings, or grid lines or spacing to distinguish or separate the headings from the data. Within the data cells, align text to the left and align numbers centrally or by decimal points. Decide whether to use grid lines or borders between columns and between rows. Basically, aim for an appearance that is helpful to the reader but is not cluttered.

Every table must have a table number and a caption/title. The numbering style must match the system you are using for the illustrations but remember that they have their own sequence of numbers. As with figures, either number tables sequentially through the whole report or use the section number followed by the table number within that section: Table 8 or Table 3.2. Tables in appendices have their own set of numbers preceded by the letter of the appendix: Table A3, Table B2. Write the word *Table* in full starting with a capital letter and write the number as a numeral, not a word. Put a full stop after the table number.

For example:

Table 12. Sea ice loss in the 20th century.

As with illustrations, captions are names for the tables so prefer to use short phrases rather than full sentences. Again, consider using sub-captions or subtitles if you want to add some descriptive text and write them in italics or put them in round brackets. Match the punctuation and writing styles to those you are using for the illustrations.

For example:

Table 12. Sea ice loss in the 20th century. *There is no explanation yet for the dip between 1910 and 1930.*

Unlike with figures, by convention place table numbers and captions above tables because tables can overflow onto the next page, which figures do not do. If a table does overflow, repeat the column headings on the next page. Align the captions either left or centre following the precedent you set for the illustrations and place tables close to the text that discusses them.

Finally, as with illustrations, you have more freedom to use abbreviations (including &) in tables than in text because space is limited, but do not use them unnecessarily.

See Chapter 6 for some examples of tables.

4.12 – Numbers and SI units

There is a convention for writing numbers especially in technical texts, although many do not seem to be aware of it: use words for numbers from zero to nine or ten (three, seven, nine) and use numerals for higher numbers (11, 12, 23). Some style guides say ten should be a word, others say it should be a numeral – your choice. (Using numerals for all numbers can pose problems as 'zero' and 'one' can be used in figures of speech, and the numeral '1' can look like the lower-case letter 'l' or a capital 'I' in some fonts.)

However, there are exceptions:

- Use numerals for all numbers in the same sentence if even one of them is greater than nine (or ten): 'between 7 and 14', not 'between seven and 14'. In some cases, you may choose to apply this to all numbers in a paragraph or even throughout the main body of a technical report if there are lots of numbers. As usual, apply your decision consistently.
- Use numerals for all numbers in illustrations, graphs, tables, etc. as you have little space.
- Use words and numerals to distinguish between two types of numbers when used close together: twelve 240-volt sockets, 16 three-pin plugs, fifteen 40-pin DIL sockets.
- Use numerals with decimals, fractions, percentages, figure and table numbers, page numbers, etc. For example: 3.5, 3½, 5 %, 25 %, Figure 3, page 4.
- Write out numbers when they start a sentence: Thirty-three tests were carried out. Or you could rephrase: There were 33 tests.
- Always use a leading zero for decimal numbers between minus one and plus one.
- Use numerals for measurements: 8 kg, 2 cm, 40 °C.
- Use numerals for the time of day but prefer the 24-hour clock, as in 14.15. If you use the 12-hour clock, write a.m. and p.m. in lower case letters with full stops as in 2.15 p.m. That appears to be the most recommended style although some American writing prefers capital letters and some writers prefer to omit the full stops.
- Decide how to write dates. Within text, the British style (24 January 2020) is more logical than the American style (January 24, 2020) but use the style best suited to your readers. Prefer to use a word for the month to avoid confusion between the British and American styles for all-number dates (11/12 or 12/11 for 11 December). The month may be abbreviated or written in full. Also, the YYYY-MM-DD system is available as an international standard, for example 2021-01-24 for the 24th January 2021.
- Use million not millions, so seven million or 7 million, except in phrases such as several millions.

(Note: The word 'billion' is now accepted to mean 10^9. Its historic meaning in British English was 10^{12}.)

SI conventions

The International System of Units (SI) was established in 1960 but its roots go back to the Metre Convention of 1875. As well as defining units, it provides conventions for writing numbers and units. The system has been revised many times and the latest revision was in 2019.

SI unit symbols are always written in the singular, so kg not kgs, V not Vs, etc. A symbol is not an abbreviation so do not use a full stop after a symbol unless it ends the sentence, so 10 kg not 10 kg. and so on.

The SI system requires a space between the number and the unit symbol. The only exceptions are for the unit symbols for degrees, minutes and seconds for plane angles where no space is required, as in rotating something by 45°. A space is also required when the value is used as an adjective, as in a 30 cm square. If you write this in words then the normal rules of grammar apply, as in a thirty-centimetre square.

The SI system applies this rule of a space between the number and the symbol to degrees Celsius and percentages: 20 °C and 5 %. However, many non-technical style guides and some technical ones say to omit the space: 20°C and 5%. I suspect that many technical report writers would agree as the text looks neater that way. Whatever you decide, be consistent but be aware that the SI system is clear that there should be a space.

Use a full stop for the decimal point (decimal marker) but be aware that the SI system also permits using a comma as the decimal point and many countries do use the comma in that way.

The SI system allows numbers with many digits to be split into groups of three for easy reading, but that is not a requirement. Engineering drawings sometimes run all the digits together rather than use spaces. When using groups of three, which is common in technical reports, SI insists that you separate them with a space, as in '30 000', not a comma, as in '30,000' (presumably to avoid confusion in countries where they use a comma as a decimal point.) Use a comma if it is your local custom but change to spaces if writing for an international readership. Do not separate the digits if there are only four of them before the decimal point: 2500.

Finally, conventionally you should write variables in italic type and symbols in roman type (non-italic) as in 'x kg'. Few report writers follow that rule although an editor might change it if the report were to be published.

4.13 – Chapter Summary

In the introduction to this chapter, I used some quotations about writing from eminent writers of bygone years and I shall end the chapter with two more as a summary.

> '*Unlike medicine or the other sciences, writing has no new discoveries to spring on us. We're in no danger of reading in our morning newspaper that a breakthrough has been made in how to write a clear English sentence – that information has been around since the King James Bible. We know that verbs have more vigor than nouns, that active verbs are better than passive verbs, that short words and sentences are easier to read than long ones, that concrete details are easier to process than vague abstractions.*'

That was written in 1976 in one of the classic American books about writing: *On Writing Well* by William Zinsser (1922-2015). It is now in its seventh edition. Bear in mind that the King James Bible was published in 1611, more than 400 years ago.

George Orwell (1903-1950), a famous British author, offered this advice in 1946 in his now famous six rules for writing (*Politics and the English Language*).

1. '*Never use a metaphor, simile or other figure of speech which you are used to seeing in print.*
2. '*Never use a long word where a short one will do.*
3. '*If it is possible to cut out a word, always cut it out.*
4. '*Never use a foreign phrase, a scientific word or a jargon word if you can think of an everyday English equivalent.*
5. '*Never use the passive where you can use the active.*
6. '*Break any of these rules sooner than say anything outright barbarous.*'

5

EDITING AND PROOFREADING

Earlier, we said there are three stages to writing a technical report: planning, writing and editing. Although you have now written the report you are not yet finished. Your good planning will have saved time during the writing stage and now you should invest some of that saved time to polish your report. Editing is like turning a rough diamond into a gem, which means there is a lot of cutting and polishing to do before you can be happy.

This is what William Zinsser had to say about the need for effective editing.

> *'Few people realize how badly they write. Nobody has shown them how much excess and murkiness has crept into their style and how it obstructs what they are trying to say. If you give me an eight-page article and I tell you to cut it to four pages, you'll howl and say it can't be done. Then you'll go home and do it, and it will be much better. After that comes the hard part: cutting it to three.'*

Many professional authors rewrite their documents several times. With each rewrite they cut and improve, moving closer to what they will finally accept. You will have neither the time nor the inclination to do that, and that is right and proper. However, you should edit to improve your report to the point where it is fit for purpose. Even a slightly flawed gem is better than a rough diamond.

Take a break between writing and editing if you can. Writing is about creating whereas editing is about criticising and improving, and that benefits from a fresh mind. If you start editing too soon, it can be hard to spot things that need improving because the writing is still fresh in your mind. Leave the report for a day or two before editing, longer if you can. Of course you will have a deadline looming, but take a break if possible.

There is no single correct way to edit a report but I suggest you start by printing and reading your entire report and make pencilled comments without using the grammar checker, and then use the grammar checker afterwards. Of course, you can just as well do it the other way round, using the grammar checker first and carefully reading a printed copy afterwards. It is surprising how many people do not read their own reports, but you cannot edit your report if you do not read it. You may find it easier to read and edit your report on paper than on screen.

As a learning exercise, read a report you wrote two or three months ago. Now that your memory of it has faded it is much easier to see how you could have improved it, and that learning may help you to improve your new report as you edit it. When I was writing regular articles for magazines I would sometimes wince when I read my articles in print, even though I was happy with them when I submitted them several months earlier.

When you turn on the grammar checker, let it check your report section by section looking carefully at what it tells you. Use it sensibly and use it for advice. Do not automatically accept everything it tells you but do think about its messages. If you disagree with them, reconsider carefully what they are saying – maybe you will change your mind.

While editing, keep asking yourself these three questions even as you finish and the report is up to the standard you want:

- Is the meaning clear?
- Is the writing concise?
- Is the report correct, both in terms of content and the use of English?

When you believe it meets these three criteria, give it to someone else to review. Your final task will be to proofread it, again after a break if you can manage that. (I know, too little time and you are already late.)

5.1 – The Grammar Checker and Readability Statistics

The Microsoft grammar checker is often accused of getting things wrong, which it sometimes does. However, it is programmed with more information about English grammar and style than most of us have ever mastered and it is worth learning to use it effectively. The version I have as I write (2020) can check for 67 grammatical issues. Unfortunately, Microsoft have a habit of changing its presentation with each new version of Word, but the actual checker seems to remain consistent and has probably improved (although that is a gut feeling and not something I have studied). Above all, treat it as a guide not as an infallible dictator.

However, change two settings before letting it loose on your report. First, within the 'Options' under writing style select whichever style checks the most grammatical issues (the issues are listed under the writing style's settings). Some styles only check at a basic level and you may think you have checked everything when you have not. You may have a choice of several styles or only two depending on which version of Word you are using. If it offers a style attractively called 'Technical', ignore it because it ignores passive verbs amongst other things. You want the style that checks the most issues, which may be called 'Formal' or 'Grammar and Refinements'.

Second, tick the check box marked, 'Show readability statistics.' The default setting is 'off' but once you have ticked it then Word will calculate some useful statistics about your writing. It does this immediately after completing a spelling and grammar check. Without ticking that box, you will never see the statistics and may never know they exist. We will look at what they tell you in a moment.

Here are some of the grammar checker's messages that seem to be the most common in technical reports.

- Passive voice (some versions say 'Consider using active voice' or 'Saying who or what did the action would be clearer') – you know all about this one and should look carefully at what it is highlighting. Be wary of automatically accepting the recommended replacement as it is not always a good choice. Although Word usually offers good alternatives there are times when it is better to write your own, and times when you should keep the passive.

- Nominalisation – this is included in recent versions of Word, otherwise use the Edit/Find function to look for the word suffixes we discussed earlier.

- Subject-verb agreement – this sounds like a compliment, but it is not. It suggests that you have muddled up singulars and plurals, maybe a singular subject with a plural verb or vice versa.

- Fragment – meaning this is not a complete sentence. It may wrongly highlight the last short-style bullet point if it starts with a lower-case letter and ends with a full stop.

- Capitalization – with the American spelling. You have a lower-case letter where there should be a capital. Note that Word now tries to put a capital letter at the start of all bullet points, which can be irritating.

- Long sentence – this will be true but not hugely helpful as it only triggers when the sentence has over 60 words. Very occasionally you might need such a long sentence but almost all your sentences should be considerably shorter than that.

- That or which – Word is usually right if it suggests that you have used the wrong word. The distinction is clearer in American English than in

British English but is correct in both, although British English allows more freedom. See the style guide in Chapter 8.

Readability statistics

As mentioned earlier, as well as checking grammar and style Word will also report some statistics about your writing, the most useful being the average length of sentences (in words) and the percentage of sentences that are passive. In the default setting these readability statistics are switched off so you need to turn them on, or you will never see them. Do this under Tools, Spelling and Grammar, Options, Show readability statistics. It will display the statistics in a pop-up box after completing a spelling and grammar check.

You can highlight a group of words, such as a section or subsection, and get the statistics for that section only. That is helpful as we would like slightly different results for different parts of the report.

The readability statistics appear in three groups that are counts, averages and readability.

It counts the number of words, characters, paragraphs and sentences. The number of words is the most useful count as you can compare before and after editing. Expect to cut around 10 % of the words from your draft. You could set that as your goal for conciseness, although it does depend on how wordy your natural writing style is.

For averages, it calculates the average number of sentences per paragraph, the average number of words per sentence and the average number of characters per word. As you already know, the average sentence length is the one to watch. You will want this to vary a little between sections while aiming for a target of around 17 to 22 words across the whole report, a little lower in the widely read sections and a little higher in the heavy technical sections.

It gives three readability statistics: the Flesch Reading Ease, the Flesch-Kincaid Grade and the percentage of sentences that are passive. Two of these are important.

The percentage of passive sentences you already know about. Aim for something around 25 % to 45 % overall, with a lower percentage in the widely read sections and a higher percentage in the heavy technical sections. Ideally, the executive summary should have the fewest passive sentences, perhaps 20 % or less.

The Flesch Reading Ease attempts to measure how easy a passage is to understand and is a standard measure that has been around for over 70 years. It applies the average length of words (measured in syllables) and the average length of sentences (measured in words) to a standard formula and gives the result as a number. It is not a percentage. The higher the score, the easier the passage is to understand.

As a rough guide, the scores mean:

- Less than 30 – very difficult (say postgraduate level)
- 30-50 – difficult (say undergraduate level)
- 50-60 – fairly difficult
- 60-70 – standard
- 70-80 – fairly easy
- 80-90 – easy
- Over 90 – very easy.

Bear in mind that these categories are guides only and are based on texts read across the entire population.

I once ran a few simple tests on some British newspapers, which I will class here as either tabloids or broadsheets. Very roughly, there was a difference between the two of about ten points, the tabloids scoring fairly easy to standard and the broadsheets scoring standard to fairly difficult. I could also measure a small difference between sections within the same newspaper. As you might expect, the finance section was the hardest to read and the sports section the easiest (which reflects my suggestion to change slightly your style between the heavy technical and the more general parts of your report). The result is reassuring as it demonstrates that the reading ease is a meaningful measure.

Remember that the Flesch Reading Ease is a standard test and is based on just two measures, the average number of syllables in the words and the average number of words in the sentences. In other words, short sentences with short words are easier to read than long sentences with long words. I suppose that is blindingly obvious, but now you can measure it and that gives you a helpful benchmark.

What score should you look for? This will depend on the types of technical report you write. With internal reports you can ask your readers what seems to be about right. With external reports you will have to use your judgement. Aim for an overall score of around 45 to 60 but a bit higher in the executive summary, conclusions and recommendations – maybe up to 65, and a bit lower in the heavy technical parts – maybe dipping down into the 30s. These can only be rough guidelines. The average score for this book is 59.5, which puts it on the borderline between standard and fairly difficult – perfect light reading for engineers.

The Flesch-Kincaid Grade is similar to the reading ease except that it equates to the education grades in the American school system. It is more applicable to the education sector than to engineering. Use the reading ease score instead.

5.2 – Reviewing

Any report that goes outside the organisation must be reviewed by a competent reviewer. An author cannot make an unbiassed judgment because they know the report far too well. If you wrote something, how can you read it as someone seeing it for the first time? How can you get that vital first impression?

Reviewers perform two distinct roles.

First, reviewers should check that the technical message is robust, that it meets the report's aims and is suitable for the intended readers. This includes checking the structural layout of the report, probably based on a given template; the technical accuracy and completeness; and whether there are any inconsistencies or contradictions in the message, its logic or argument. Depending on the organisation, reviewers can either advise or insist on changes.

Second, reviewers should check the language and presentation of the report to ensure they are clear, concise and correct, and suitable for the intended readers. This includes ensuring that the message is well written and flows well, and that it uses plain English as much as possible with good spelling, grammar and punctuation. It also includes checking that style decisions are applied consistently and agree with the organisation's template and style guide, if there is one. It is not their job to rewrite parts of the report in their own style because they prefer it to yours, but they are expected to advise.

One reviewer may review everything or the reviewing may be split between two people, one checking the technical content and the other checking the language and style. Overall, they need to ensure that the report is fit for purpose and fit to represent the organisation.

Normally, they do not proofread as that is your job.

5.3 – Proofreading

You have now decided that your report is technically robust and that the English is clear, concise and correct, and your reviewer agrees – maybe after some changes. Everyone is happy and now comes the final quality check, the proofreading, which you do. It is safe to say that proofreading always finds things to correct but by now those things should be minor. Proofreading is laborious and can be tedious. Unlike editing, where you expect to make significant changes and do some rewriting, proofreading is the final check for mistakes or inconsistency in the spelling, grammar, punctuation and formatting.

Do as good a job as you can within a sensible time frame. Do not worry if you miss a few things, especially typing errors which are hard to spot if they result in a real word, like typing 'of' instead of 'if'. Even books written by professional writers and published by famous publishing houses contain a few small mistakes.

Many people find it easier to proofread on paper than on a screen. Use a ruler or a piece of paper to cover most of the page (whether on paper or screen) to help you to limit your attention to one line at a time. This slows you down and helps you to spot mistakes or inconsistencies.

Be methodical and keep hand-written notes of decisions you make to help you to be consistent with such things as spellings, hyphenated words, caption style and so on.

Here is a short list of some of the consistency checks:

- font style and size
- line spacing and alignment
- spaces between sentences, prefer one space
- page numbering
- headings and subheadings
- figure and table numbers and captions/titles
- citations to figures and references
- style used for references
- bullet point punctuation and phrasing
- headers and footers
- hyphenated words
- use of numbers, scientific units and symbols
- acronyms spelt out on first and maybe subsequent use
- spelling where there are alternative spellings
- spelling of names
- date and time formats
- use of i.e. and e.g.

5.4 – Chapter Summary

Good planning should lead to better writing and easier editing – and save time. Take a break between writing and editing as you try to approach this criticising task with a fresh mind. This does not work perfectly but you will get some benefit from it, time permitting.

First, print and read the report and pencil in any corrections. Second, with the spelling and grammar checker turned on, edit your report carefully section by section on the computer. Select the writing style that checks the most grammatical points, there are nearly 70 of them. Examine carefully what the grammar checker tells you but make most of the decisions and changes yourself.

Turn on the readability statistics and check the average sentence length, the percentage of passives and the Flesch reading ease score for each section or subsection. Listen to your reviewer's advice and make any further changes that are necessary. Proofread carefully, despite the boredom. Then wish the report well, send it on its way and get on with the rest of your work.

6

SI UNITS

Formally known as the Système International d'Unités (SI), this is the internationally agreed system for units. The SI was established in 1960 but its roots go back to the Metre Convention held in Paris in 1875. It has been revised many times. The details below are from the 2019 revision. The system has seven base units, twenty-two derived units with what are quaintly called 'special names and symbols', some acceptable non-SI units and a set of prefixes. These are displayed in Tables 1 to 4 below.

Symbols that are named after people start with a capital letter; the rest start with a lower-case letter. There is one exception to this rule. The symbol for the litre can be either a lower-case 'l' or a capital 'L' because in some fonts the lower-case 'l' is identical to the numeral '1'.

Table 1. SI Base Units

Quantity	Unit Name	Unit Symbol
Time	second	s
Length	metre	m
Mass	kilogram	kg
Electric current	ampere	A
Thermodynamic	kelvin	K

temperature		
Amount of substance	mole	mol
Luminous intensity	candela	cd

Notes:
The kilogram is the only unit with a prefix.
The American spelling of metre is 'meter'.

Table 2. SI Derived Units

Quantity	Unit name	Unit symbol
Plane angle	radian	rad
Solid angle	steradian	sr
Frequency	hertz	Hz
Force	newton	N
Pressure, stress	pascal	Pa
Energy, work, amount of heat	joule	J
Power, radiant flux	watt	W
Electric charge	coulomb	C
Electric potential difference	volt	V
Capacitance	farad	F
Electric resistance	ohm	Ω
Electric conductance	siemens	S
Magnetic flux	weber	Wb

Magnetic flux density	tesla	T
Inductance	henry	H
Celsius temperature	degree Celsius	°C
Luminous flux	lumen	lm
Illuminance	lux	lx
Activity referred to a radionuclide	becquerel	Bq
Absorbed dose	gray	Gy
Dose equivalent	sievert	Sv
Catalytic activity	katal	kat

Table 3. Acceptable Non-SI Units

Quantity	Unit Name	Unit Symbol
Time	minute	min
Time	hour	h
Time	day	d
Length	astronomical unit	au
Plane and phase angle	degree	°
Plane and phase angle	minute	'
Plane and phase angle	second	"
Area	hectare	ha
Volume	litre	L or l

Mass	tonne/metric ton	t
Mass	dalton	Da
Energy	electronvolt	eV
Logarithmic ratio quantity	neper	Np
Logarithmic ratio quantity	bel	B
Logarithmic ratio quantity	decibel	dB

Note: The symbol for litre can be either a capital L or lower-case l because in some fonts the lower-case l is identical to the numeral one. Also note that the following units are not included in the 2019 edition of the SI Units: nautical mile (M), knot (kn), bar (bar) and the angstrom (Å).

Table 4. SI Prefixes

Factor	**Name**	**Symbol**
10^1	deca	da
10^2	hecto	h
10^3	kilo	k
10^6	mega	M
10^9	giga	G
10^{12}	tera	T
10^{15}	peta	P
10^{18}	exa	E
10^{21}	zetta	Z

10^{24}	yotta	Y

10^{-1}	deci	d
10^{-2}	centi	c
10^{-3}	milli	m
10^{-6}	micro	μ
10^{-9}	nano	n
10^{-12}	pico	p
10^{-15}	femto	f
10^{-18}	atto	a
10^{-21}	zepto	z
10^{-24}	yocto	y

Write the prefixes without a space between the prefix and the unit symbol (so kV not k V) and both should be in roman type (not italics). Most of the multiples use capital letters whilst all the submultiples use lower-case letters.

Note: The SI system specifies these prefixes for multiples of ten. In the computer industry, and in the wider public domain, they are also used as prefixes for multiples of two, as in one kilobyte representing 1024 bytes, not 1000 bytes. The SI system recommends another set of prefixes for multiples of two. The names of these other prefixes are the first two letters of the regular prefixes plus the letters *bi*, so for instance *kilo* becomes *kibi*. The symbols are the normal prefix letter, but always as a capital, followed by the letter *i*, so *k* becomes *Ki*: kibi (Ki), mebi (Mi), gibi (Gi), tebi (Ti) and so on.

7

BOOK SUMMARY

Here is a summary of the main advice given in this book. A good technical report meets many criteria, but it will not do this without careful thought and hard work. If the writer does not write clearly, then the readers will struggle. Easy writing means hard reading.

There are three phases to producing a really good technical report. They are: planning, writing and editing. Treat each one as a separate activity.

Planning is the key to a well-organised report (and poor organisation is probably the biggest moan about technical reports). Clear and concise writing is the key to explaining things to your readers, but keep asking yourself what is the best way to explain this or that point to them. Finally, editing is the key to making it as good as you sensibly can and turning your rough diamond into a sparkling gem.

7.1 – The Fundamentals

1. Many people do not read the entire report but focus on the parts that matter most to them. In a way, that is good news because it allows you to adjust your writing style, such as choice of words, for the expected readership. Almost everyone will read the executive summary, and most people will read the aims, background, discussion, conclusions and the recommendations. Engineers will read the results and analysis, perhaps partly on behalf of their managers. The discussion will pick up more readers as non-specialists seek to understand what it all means.

2. Aim or purpose: Do not start without a clearly defined aim or purpose, maybe composed of several distinct objectives. The aim or aims may be specific questions which will have specific answers. The aims pose the questions and the conclusions and recommendations answer them, even if the questions are not explicitly phrased as questions. The aims may be given to you as part of a terms of reference.

3. Ethos: the report should demonstrate the credibility and authority of the writer and his or her organisation to answer the questions posed.

4. Logos: the report should have a logical structure. Sequence it so that your readers can follow the reasoning easily. The intended readers will notice and appreciate when the information progresses logically from the beginning to the end – and they will become frustrated when it does not. It should be comprehensive but not overburdened with unnecessary details, and it should report honestly contrary evidence and explain why it can be discounted, if it can be.

5. Pathos: the report should be aimed at the intended readers. It meets their needs yet allows the writer to add other points that he or she believes the readers ought to be aware of.

6. The report should be technically robust and contain the right amount of relevant detail, that is, enough to derive the conclusions. Additional details that some readers may find interesting, but are not needed to reach the conclusions, can be in an appendix.

7. Overall, the report should be clear, concise and correct.

7.2 – Planning

8. A lot of what makes a report a good report lies in the sequencing of the information. It is much better and easier to design the sequence before writing than while writing. A good sequence has to be designed, not left to chance.

9. Start your sequence at Level 1, the main sections, which are usually given in your organisation's template. If there is no template, then use the conventional format suggested in Chapter 3, which is: the Executive Summary; Introduction, Results, Analysis, Discussion, Conclusions, Recommendations, References, and Appendices if needed. Modify the format as necessary bearing in mind that very few reports will need every item mentioned in Chapter 3.

10. Then move on to Level 2, the subsections, and Level 3, the paragraphs. Use bullet points (or mind maps) for the main items of information and click and drag them until you get the sequence you want. This gives you a plan or storyboard for the report. You choose how detailed this plan will be but making it considerably more detailed than you are used to will almost certainly give you a better report and save time.

11. Executive Summary: Almost everyone reads this. It must be able to stand alone without the report (some readers may only receive the executive summary) and it must be self-contained (not needing readers to look in the report to understand something). It poses and answers the question and it should be short. It gives enough of the background, methods, results and reasoning to put the question and answer into context, giving the readers confidence in the report. Outlining the scale

of the results may be a good idea.

12. Introduction: This usually starts with the background followed by the aims – or sometimes the aims followed by the background. Then come the methods, which may be a subsection of the introduction with their own subheading. If the methods are complicated then they may deserve a separate section, which would come next – before the results.

13. Results: Include the right amount of relevant detail, everything that is needed to reach the conclusions (perhaps after analysis and discussion). Any additional data that some of your readers might find interesting could go into an appendix. The results should be purely factual and dispassionate. In many reports the results section will have several subsections, each for different sets of results.

14. Analysis: If justified, include an analysis of the results to bring out information derived from them. If little analysis is needed then perhaps you could include it in the discussion.

15. Discussion: This should explain what it all means, especially for the intended readers. It gives your considered professional opinion or opinions. If there are competing options, discuss their relative merits. Do not brush aside any conflicting results but instead consider them and explain why you have chosen to set some aside, if indeed you have. As non-technical readers are likely to read it, the discussion should be more understandable to a wider readership than the results and analysis so reduce the technical jargon. Non-technical readers may not want all the gory details, but they do want to know what it all means – the implications for them. Sometimes, the first few paragraphs can be aimed at them, with zero jargon, before moving on to a more complicated discussion for the specialists. Many of the conclusions may arise naturally as part of the discussion and that is fine.

16. Conclusions: All the conclusions should be fully supported by and derived from the evidence presented. Let the conclusions arise naturally within the discussion if they do, but then present them boldly, concisely and in plain English in the conclusions section. Give each its own paragraph, although sometimes bullet points will suffice. Many non-specialists will read the conclusions and they will see them here for the first time; others will already be familiar with them but will want to see them all in one place.

17. Recommendations: These too should be bold and concise, each with its own paragraph or the whole as a set of bullet points. Many non-specialists will read them so they must be clear and precise. Write them in plain English with as little jargon as you can manage.

18. References: If you have references, decide whether to use the Harvard or Vancouver system. The Harvard system cites references in the text using the main author's name in brackets with the year of publication

and lists them all at the end in alphabetical order by the main author's name. The Vancouver system uses numbers to cite references and lists all the references at the end in numerical order.

19. Appendices: Use these for additional data that is not needed to reach the conclusions but that some readers may justifiably be interested in seeing. Appendices are identified by letters: A, B, C, etc. Annexes are written by others and are included for the convenience of the readers.

20. Number the sections and subsections of the report. Try to limit yourself to a maximum of three layers, such as subsections 1.3.1, 2.4.2 and 3.6.3. Use letters for appendices, numbering sections as A1.1, A1.2, A1.3, etc.

21. Managerial format. Only use this format if asked to do so. Its focus is to immediately provide the question and answer, mainly for non-technical managers, followed by the details. The format contains the usual sections but rearranged as: the Executive Summary, Introduction (including the aims), Conclusions, Recommendations, Method, Results, Analysis and Discussion.

22. 1:3:25 format. Only use this format if asked to do so. There are three sections, nominally of one page, three pages and twenty-five pages. The one page is for the main or key messages and can be presented as a long set of bullet points after a very brief introduction. It answers the 'So what?' question and may not mention the methods or results. The three pages are an Executive Summary but it is longer and more detailed than the traditional approach, being about 10 % of the whole. Lead with what matters most to the readers. The main body of around 25 pages has up to seven sections: the context, the implications, the approach, the results, additional resources or information, suggested further research, and the references or bibliography. More than 25 pages may be needed.

7.3 – Writing

23. Aim for good grammar, punctuation and spelling.
24. Try to avoid interruptions when writing, and that includes turning off the spelling and grammar checkers.
25. Try to achieve writing that is good enough not to distract your readers from your message; it should not trouble them. In a sense, the grammar, spelling and punctuation should be transparent to your readers. Put their understanding first.
26. Aim for a clear, flowing style that uses words the readers understand. You want them to understand it all at the first attempt – which is difficult to achieve but try to get as close to that as you can.
27. Remember to keep paragraphs and sentences short, and use plain words as much as possible.

28. Think of an average of four to eight lines per paragraph or four to eight paragraphs per page (A4 or US letter).

29. Aim for an average sentence length of about 17 to 22 words over the entire report. But aim for a lower average length of about 15 to 20 words in the widely read sections: the executive summary, discussion, conclusions and recommendations. Accept a slightly higher average, about 20 to 25 words, in the heavy technical parts of your report.

30. It is the average sentence length that matters so use longer sentences when you need to explain complex points. However, avoid extremely long sentences. (The Microsoft grammar checker does not flag a long sentence until it reaches 60 words, which is extremely long.)

31. Structure your sentences in a straightforward way, often starting with the subject followed by the verb and then the object.

32. Take care when putting asides in the middle of sentences. Where sensible to do so, prefer to put them either at the beginning of sentences or at the end.

33. Be specific, not vague.

34. Generally, prefer active verbs to passives ones. As engineers, our education has trained us to use too many passive sentences so try to convert many of them into active ones. However, it is not a case of actives are good and passives are bad, you will need more passive sentences in the technical sections than in the more general sections. Remember that changing a passive into an active may change the focus of the sentence, which you may not want to do. Preferring actives will shorten your report and make it a bit more direct. However, using more passives can make it appear more formal, which some may see as more authoritative. It is a question of degree.

35. Nominalisation: Where possible, prefer a verb such as 'sterilise' to a noun such as 'sterilisation'. Preferring verbs will shorten the report and make it a bit more direct. Again, our education has led us to overuse nominalisations. Recent versions of the Microsoft grammar checker will find nominalised verbs (nouns) which might be suitable for conversion into real verbs. Once again, it is not a case of nouns are bad and verbs are good; you will need both. It is a question of degree. In some versions of Word, the grammar checker does not flag nominalisations. Use the Edit/Find function to search for words ending in *-tion*, *-ment*, *-ance*, *-ence* and *-ity*, especially those followed by *of* and preceded by *the*, *a* or *an*.

36. Use plain English where you can. Your readers in the main technical parts of your report are likely to be technically qualified and should understand the normal jargon. In the widely read sections try to avoid jargon where you can. Do not seek 'fancy' words to show off. Generally speaking, readers do not appreciate it.

37. Spell out any acronyms on first use, and again if you use them in the discussion, conclusions or recommendations. Avoid all acronyms in the executive summary if possible.
38. Use emphasis sparingly and prefer italics to bold. Only use underline for hyperlinks. Prefer to use careful phrasing to emphasise points.
39. Use lists, illustrations and tables where they work better than prose.
40. Every illustration and table should be numbered and have a caption (a separate series of numbers for illustrations and tables). Either number them through the report or by chapter. Every figure and table must be mentioned in the text. See the text as either leading or supporting an illustration or table. One takes the lead in giving the message while the other supports it. Appendices have their own numbering system preceded by the letter of the appendix.
41. Do not nitpick over minor stylistic issues, you have better things to do.
42. Use parallelism or symmetry as appropriate, especially in lists such as bullet points.
43. Consider punctuation to be a service to your readers to help them to understand your writing – not as a set of half-remembered rules to please a professor of English language. Be sensible, obey the main rules that do exist but maintain some flexibility as you seek to ensure clarity for your readers. Be especially careful and use a lot of common sense with commas, hyphens and dashes – see the style guide.

7.4 – Editing

44. Take a break between writing and editing as you try to approach this criticising task with a fresh mind. This does not work perfectly but you will get some benefit from it.
45. First, print and read the report and pencil in any corrections. Then, with the spelling and grammar checker turned on, edit your report carefully section by section on the computer. (Of course, you can just as well check with the grammar checker first and then check on paper afterwards.) In the grammar checker, select the writing style that checks the most grammatical points, there are nearly 70 of them. If a style called 'Technical' is offered, reject it in favour of 'Formal' which checks more grammar. In more recent versions of Word, chose 'Grammar and Refinements'. Examine carefully what the grammar checker tells you but make most of the changes yourself.
46. Every writer uses too many words in the draft. Remember the often-repeated advice: Cut! Cut! Cut!
47. Turn on the readability statistics and check the average sentence length, the percentage of passives and the Flesch reading ease score for each section or subsection. Listen to your reviewer's advice and make any further changes that are necessary.

48. Check the minor style issues for consistency. There are many of them including: font and font style, line spacing and alignment, spaces between sentences (prefer a single space), headings and subheadings, figure and table numbers and captions, bullet point style, hyphenated words, use of numbers with scientific units and symbols, and so on.
49. Proofread carefully, despite the boredom. Then wish the report well, send it on its way and get on with the rest of your work.
50. Finally, make sure your report is presented on time.

8

STYLE GUIDE

Many elements of style are neither right nor wrong but depend on personal choice and I have consulted many style guides while writing this one. Generally, I have turned to the Oxford University Press publications, *New Hart's Rules, the Oxford Style Guide* and the *Concise Oxford Dictionary*. I have also checked with several online sources including the *Cambridge* and *Merriam-Webster* dictionaries, the *Chicago Manual of Style* and several other style guides. The bibliography lists the guides I have consulted most.

There is a broad agreement between the main style guides on the most important issues although they may not agree on all of the less important points – that is the nature of style, and style changes with time.

If your organisation has its own style guide then, of course, you must follow it although very occasionally a client may specify points of style for you. If neither of these applies, then take the guidance offered here. A style guide could go on for ever so if you need guidance on something that is not covered here then check with several online sources and make your own decision. Unless you choose something completely wrong, consistency is sometimes more important than exactly what you do, especially with minor points of style. As always, your organisation's or your client's decisions take precedence over any other advice.

1:3:25

This is a report format that originated in Canada and is sometimes used for medium length reports of about 30 pages. It makes the executive summary longer and more detailed than usual, nominally 3 pages and precedes it with a short set of key points, often presented as bullet points on a single page, and follows it with the main report of around 25 pages. The numbers are strong guides but not rules. This format is unusual so only use it if you are required to do so.

See also: Section 2.10 in Chapter 2.

A, an

Use *a* before words (including acronyms) that sound as if they begin with a consonant, irrespective of whether they do so or not. Use *an* before words that sound as if they start with a vowel, irrespective of whether they do so or not. For example: a university (you-ni-ver-sity), a cubicle, a test tube, a DC supply, a UV test (you-vee), an hour (silent 'h'), an element, an industrial setting and an AC supply.

Abbreviations

Abbreviations are words that have been shortened by omitting the end, for instance *approx.*, *dept.* and *misc.* Most abbreviations end with a full stop unless they are everyday abbreviations, such as *Prof*, *Tues* and *Nov.* Ideally, do not use any abbreviations in reports, especially in the executive summary, conclusions and recommendations.

See also: Acronyms, contractions.

Absorb, adsorb

One letter makes all the difference. *Absorb* means to take something in – it is internal. Plants absorb carbon dioxide, children absorb information. *Adsorb* means that something sticks to the surface – it is external, like gas molecules sticking to the surface of a solid.

Accuracy, precision

These do not mean the same thing. *Accuracy* refers to how close a measured value is to the true value. *Precision* refers to how close to each other several measurements of the same thing are. You could have precise measurements (they are all close to each other) but they are not accurate (not close to the real value) because the meter was not calibrated properly. Also, you could have several measurements of the same thing that are not precise (the values are scattered) but their average value might be accurate (close to the real value).

Acronyms, initialisms

Acronyms are terms that you pronounce as a word, such as radar, sonar and scuba, whereas *initialisms* are terms that you pronounce as a series of letters, such as PDF, DVD and UHF. Some terms combine the two approaches as in JPEG (pronounced J-peg). Most acronyms are written in lower-case letters while most initialisms are written in capitals without full stops (US: periods). In this book, other than in this style guide, I have used *acronym* as a general term for both as few engineers distinguish between them.

Ban initialisms from the executive summary if you can and use as few as possible in the conclusions and recommendations. As general guidance for the rest of the report, if you are only going to use an initialism once or twice then it is better to write it out in full each time. However, you may use a lot of technical initialisms, and use some of them frequently, in the main body. As usual, you should define them on first use and then give the initialism in round brackets, for example the Portable Document Format (PDF). If you do use initialisms in the conclusions or recommendations, then consider redefining them for those readers who skipped the main body of the report.

Plural initialisms take a small 's' without an apostrophe, for example OEMs, DVDs, MPs.

See also: Abbreviations, contractions.

Adsorb

See: Absorb.

Advice, advise

Advice is a noun; you may suggest that your client takes your advice. *Advise* is a verb; you may advise your client to follow your recommendations.

Affect, effect, impact

As a verb, *affect* means to have an influence on something or someone, or to cause something to happen, whereas *effect* means to bring about or make something happen. 'Contamination affects some results.' 'The society's ruling will effect changes in practice.'

As a noun, engineers use *effect* to mean the result of an action such as the *effect* of something happening. Of course, it is also used for physical phenomena, such as the Faraday effect and the Doppler effect. You are less likely to use *affect* as a noun because it refers to emotion or feelings.

The primary meaning of *impact* is for one body to hit or strike another as in 'The meteor impact left a crater.' *To impact* also means to have a powerful effect on something. As a noun, *impacted* means 'packed' or 'wedged in' and has a specialist meaning in some subjects, especially dentistry.

Aggravate, irritate

Aggravate means to make something worse although informally it can mean 'to annoy'. *Irritate* does mean 'to annoy'.

Alignment, justification

Many organisations prefer the text to be fully justified, which means paragraphs have straight edges both left and right. The disadvantage of full justification is that it can leave uneven spaces between some words, which can look ugly, and some say it decreases reading speed. Others prefer left justification (also called flush left or aligned left), where the left-hand side is straight and the right-hand side is jagged, which some say increases reading speed. The disadvantage of left justification is that some people do not like the jagged edge it leaves on the right.

Alternate, alternative

Alternate means to follow each other in turn or to change repeatedly between two conditions. *Alternative* refers to a choice of another option.

Among, between

Among is used for referring to things, places, choices, etc. that are not clearly separated or identified as individuals but are part of a group or mass, as in 'among many options'. Use *between* when referring to two or more specific or distinct objects, places, choices and so on, as in 'Choose between options A, B and C.'

Ampersand, &

Do not use the ampersand in the text of a report as an abbreviation for *and*. There are exceptions such as when it forms part of an organisation's name, as in 'AT&T', 'Johnson & Johnson' or 'M&S', and its use in programming languages and HTML code. Using **R&D** for 'Research and Development' is usually acceptable, and you can use it as needed in figures and tables where you are short of space.

And/or

The expression *and/or* is often used by engineers but do not use it in reports, particularly formal reports or reports about computers, programming or electronics. Instead of writing 'X and/or Y' try writing 'X or Y or both', even though it does feel a bit wordy. Clarity is vital. If you do use *and/or* then be aware that it takes a plural verb.

There are three potential problems. First, many people think it is ugly. (The *Guardian Style Guide* calls it horrible.) Second, the writer may simply mean 'or', so why not write 'or'? Third, there is an unlikely but possible objection that people who are aware of Boolean logic and the difference between the operators *OR* and *XOR* may find it confusing.

Anybody, any body, anyone, any one

In technical reports you may use *any one* in the context of 'any one of these incidents, experiments, measurements', etc. You probably will not need *anybody* or *anyone* as both mean 'any person', nor are you likely to use *any body* unless you are referring to 'any group' or to dead bodies.

Apostrophe

Possessives: Use the apostrophe to show the possessive as in 'IBM's contract'. If the singular word ends with an *s*, then use either *...s'* or *...s's* as you see fit. Both can be right although the argument over which to use can become complicated. Omitting the extra *s* is more common today, for example, 'Siemens' contract' rather than 'Siemens's contract'. One exception, which few of you will need, is when using a classical name in a scientific or engineering setting. For example, the planet Mars is named after the Roman god of war, hence 'Mars's surface', although you may prefer the 'Martian surface' or the 'surface of Mars'. Transposing words solves many difficulties.

Plurals use *...s'* as in 'the engineers' tools' for several engineers, but 'the engineer's tools' for one engineer.

Plurals of letters, words and numbers: The apostrophe may be used but this is a disputed point. Should you write 'Four hexadecimal F's' or 'Four hexadecimal Fs'? Keep things simple by not using the apostrophe and write 'Four hexadecimal Fs', 'three As and two Bs', 'the report is full of ifs and buts' and so on. An exception is when an apostrophe clarifies the meaning for the reader, such as when referring to letters or numbers as objects as in 'the counter increases by 5's' (or in that case prefer 'the counter increases by fives'). Rephrasing may be better than using an apostrophe.

Time: Although not universally agreed, use the apostrophe with time when the time period modifies a noun as in 'two days' time' and 'one week's holiday', but note: 'a three-second interval' or 'an interval of three seconds'.

Contractions: Do not use contractions such as *won't* and *doesn't* in technical reports, write the words in full: 'will not', 'does not'. The word *it's* is a contraction of 'it is' and should be written in full, whereas *its* is the possessive of 'it', which you may use in technical reports.

If you think apostrophes are a nuisance, consider the following:

- My sister's friend's suggestion = a suggestion from one friend of one sister.
- My sister's friends' suggestion = a suggestion from two or more friends of one sister.
- My sisters' friend's suggestion = a suggestion from one friend of two or more sisters.
- My sisters' friends' suggestion = a suggestion from two or more friends of two or more sisters.

Appendices, annexes

Appendices supplement the report; they are written by the same author and are referred to in the report. Use them for extra information that some of your expert readers are likely to be interested in but are not needed to reach the conclusions and recommendations. Identify appendices by letters, Appendix A, Appendix B, etc. Figure and table numbers within appendices start with the letter of the appendix, so Figure A1, etc. *Annexes* are separate self-contained documents or parts of documents and are usually written by someone else. They are included for the reader's convenience, for example a copy of a quality procedure or a manufacturer's data sheet. They, too, are identified by letters.

Average, mean, median, mode

The *average* and the *mathematical mean* are the same thing. Take a set of numbers, add them up and divide by how many there are in the set and you get the average or mathematical mean. The *median* is the mid-point or middle value of the distribution. The *mode* is the number that occurs the most in the distribution.

For example: If we have these nine numbers: 5, 9, 8, 7, 6, 10, 6, 12, 6 then the average or *mean* is 69/9, which is 7.67. The *median* is the midpoint. Rearranging the numbers in order: 5, 6, 6, 6, 7, 8, 9, 10, 12 shows the middle one to be 7, so the *median* value is 7. The *mode*, the number that occurs the most, is 6.

Because, due to, owing to

Traditionally, *because of* and *owing to* modify verbs and *due to* modifies nouns. 'The experiment *was cancelled* because of rain' or 'The experiment *was cancelled* owing to rain.' However, 'The *cancellation* of the experiment was due to rain.' If in doubt, try phrases like 'caused by', 'attributed to' or 'as a result of'.

Due to can also mean 'owing to' as in 'the money was owed to Smith', and in the sense of 'timed to' as in 'The meeting was due to finish at five o'clock.'

Between

See: Among.

Brackets, parentheses

Use round brackets, also called *parentheses*, to enclose asides in the text: 'Apply to all cases (except if fitted with Mod A)'. Prefer using commas if the asides are only weakly set aside and save round brackets for strong asides. Also use round brackets for citing figures, tables and references in the text: (see Figure 4.2), (see Table 3.5) and (Broad, 2019).

Remember that in mathematical expressions there is a preferred hierarchy of brackets – first round brackets, then square brackets and finally curly (wavy) brackets: $\{ [(\ )] \}$

Bullet point lists

See: Section 4.9 in Chapter 4.

Can, may

Can refers to the ability to do something; *may* refers to permission to do something.

Cannot, can not

Usually spelt as one word, *cannot*.

Capital letters, upper-case letters

Do not overuse capital letters. There is a tendency with some writers to capitalise any noun they feel is even half important and readers can find that irritating. Do not do it.

Obviously, use capital letters to start a sentence, including when a sentence starts with a number which must be written out as a word or words. All proper nouns start with a capital letter and that includes the names of people, specific places, companies, local authorities, legislation and so on as in Jane, London, Google, Hampshire County Council and Road Traffic Act.

The first and main words of report and book titles are usually capitalised as in *A Brief History of Time, from the Big Bang to Black Holes*.

All the symbols for chemical elements are capitalised (but no full stops), hence *Fe*, but the names of chemical elements are not, hence *iron*.

With scientific units, the symbols for units named after people are capitalised but the rest of the symbols and the names of the units are not capitalised. Hence 'W' for 'watt', which is named after James Watt, but 'm' for 'metre' ('meter' in American usage) which is named after the Greek word for measure.

Finally, the personal pronoun *I* is always capitalised, but you should not use it in a technical report except maybe in very informal reports such as email reports.

Captions

Figures and illustrations will normally fit within a page so put their captions below them. Occasionally, tables may run on to the next page so always place table captions above the table and repeat them on the next page if they overrun. Give them all a number and a title. See Sections 4.10 and 4.11

in Chapter 4 for more details.

Case

Case can be a waffle word and should be cut when it is. For example, 'There are cases when measurements were…' would be better as, 'Sometimes measurements were…'. (Both would be even better if they gave the reader either the number or percentage of times that whatever it was happened.) 'If it is the case that…' simply means 'if', and 'if it is not the case that…' means 'if not'.

Chemical elements

Write the names with lower-case letters (do not use italics) although the symbols start with a capital letter without a full stop: 'iron' and 'Fe'. There are no spaces between the element symbols in chemical formulas as in H_2O for water.

Chronic

Chronic refers to a disease or other health condition that lasts a long time. It does not mean bad or severe.

Collective nouns

Most collective nouns, such as the names of organisations or groups, are singular entities and therefore take singular verbs. 'Microsoft is introducing…' or 'The committee is deciding…'. However, some collective nouns take the plural as in 'The police are investigating…'. The guideline is to consider whether the organisation is acting as a single unit as in 'The committee is issuing a directive,' or as a group of individuals as in 'The committee are arguing about the new directive.'

Colon

Use a *colon* (:) to introduce lists, including bullets, at the end of the introductory statement, but do not use a colon if the list completes the introduction and is fully a part of the whole sentence. The word immediately following the colon is often capitalised in American style but not in British style. You will probably not need colons elsewhere if you keep your sentences short. Please note, nowadays a colon is never followed by a dash as in ':–' so do not write that.

Comma

My rather old copy of *The Concise Oxford Dictionary* comments: 'Use of the comma is more difficult to describe than other punctuation marks, and there is much variation in practice.' Too true! The *Chicago Manual of Style* lists 40 uses for the comma, proof enough that using commas can be complicated.

Use commas with discretion to help your readers; too many can be distracting, too few can make it harder for readers to understand.

Expect to use a comma after a clause or phrase that introduces the main part of the sentence, although it may not be necessary if the introduction is short. If/then statements often use a comma as in, 'If…, then…'.

Deciding whether to use commas with two or more adjectives before a noun can be tricky. Use commas if you could satisfactorily rearrange the adjectives or if you could put the word *and* between them as in 'accurate, repeatable readings'. This will also work as 'repeatable, accurate readings', as 'accurate and repeatable readings' and as 'repeatable and accurate readings'. It may not always be that simple. Try it with, 'a corroded, thin metal strip'. 'Metal' takes precedence and then 'thin' as it was thin before it became corroded. Do not use commas if you cannot rearrange the adjectives or put the word *and* between them, as in 'a tiny temperature gauge'. This cannot be rewritten as 'a temperature tiny gauge' nor as 'a tiny and temperature gauge'. Here, the final adjective takes precedence over any others and must come immediately before the noun.

Use commas to separate items in a list: 'We tested samples A, D, G and J.' In American usage, this would be: 'We tested samples A, D, G, and J.' Placing a comma before the word *and* in a list is a question of style; British English tends not to whereas American English tends to. (This is known as the Oxford or serial comma.) Use an Oxford comma for clarity where there is another *and* nearby, as in 'In order of preference, the potential suppliers are Greys, Browns, Reds and Blues, and Greens.' This shows that 'Reds and Blues' is a single organisation. Without the extra comma the organisation could be misread as 'Reds and Blues and Greens'.

Use a comma before *but* when it separates two independent, and usually contrasting, clauses in the same sentence as in, 'The samples should be clean, but two were contaminated.' How do you know they are independent clauses? Both could stand as sentences on their own. Such commas are used less in British English than American English but use one where it will help the reader.

Use commas in pairs as brackets, especially with *which*, to indicate an aside. Sometimes the start or end of the sentence may substitute for one of the commas, as in this sentence. (See also: That, which.)

The International System of Units (SI System) specifies a space, not a comma, as a separator in numbers: 23 000 rather than 23,000. Many engineers ignore this rule – it is your choice but be aware that the SI system rejects the comma because some countries use the comma as a decimal point. There should be no separation if the number has only four digits as in 4000. See also Section 4.12 in Chapter 4.

Common, frequent, regular

Common means to occur often, as in a common mistake. *Frequent* refers to something that occurs at relatively short intervals. *Regular* refers to something that happens at fixed points in time or space.

Compare to, compare with

Compare to draws attention to similarities between two things whereas *compare with* points out differences. You may compare one sample *to* another to show how similar they are, but you would compare one sample *with* another to contrast them as in comparing a damaged component *with* a new one.

Complement, compliment

Complement means something goes with, improves or completes something else. It also means the number of people or things that something normally has, as in 'the complement' of a ship's company. It has specialist meanings in biochemistry and geometry. *Compliment* is an expression of praise, approval or respect.

Comprise, consist of

Comprise means 'is composed of' or 'consists of'. *Consist* needs the word *of* whereas *comprise* does not. 'The tests comprise X, Y and Z' or 'The tests consist of X, Y and Z.' You may prefer to use *consists of* as it is probably the more common expression.

Continual, continuous

If something is *continual* then it repeats frequently but with breaks, as with dashed lines on a road. If it is *continuous* then it is uninterrupted with no breaks, as with a railway line.

Contractions

Contractions are words that have been shortened by omitting letters from the middle of the word, such as *can't*, *won't* and *it's* (meaning 'it is'). Contractions do not need a full stop at the end, but they do need an apostrophe to indicate where the missing letters should be – except for common ones, such as *Dr*, *Ltd*, and *Mrs*. Do not use any contractions in reports, except possibly in informal email reports.

See also: Abbreviations, acronyms.

Dashes

Although dashes and hyphens look similar, they are opposites in use. In general, although there are exceptions, dashes separate words while hyphens join words. Strictly speaking, hyphens are not dashes but we will not worry about that distinction.

Apart from the hyphen, there are two types of *dash*, the *en dash* and the *em dash*. You are likely to use only one of them in a technical report, probably the *en dash* in British style but the *em dash* in American style. You may have used them without realising it. The *en dash* is roughly the length of a capital N and the *em dash* is roughly the length of a capital M. So, in terms of length, the hyphen is the shortest, the *en dash* is the middle one and the *em dash* is the longest.

If you type a hyphenated word such as spin-off, which has no spaces before or after the hyphen, Word will leave it as a hyphen. If you type a hyphen with spaces on both sides – Word will convert it into the middle-length *en dash* when you write the word that follows it. To get the *em dash*, the longest of the three, type two hyphens without spaces between them and closed up to the words on both sides. In British style, you are only likely to use this—the *em dash*—if your report is published formally, even then it will depend on the publisher's style rather than yours.

To most engineers, the differences in how the three are used is probably far less interesting and less important than they are to publishers. What really matters is that our readers fully understand what we are saying. If differentiating between dashes and hyphens helps the readers, then take pains to get it right. Otherwise, let Word sort it out.

Use pairs of *en dashes* with spaces as if they are brackets (parentheses). (In the American style use the *em dash* without spaces.) You may have been doing this while thinking they are hyphens, which is fine. Just type a hyphen with a space on both sides or two hyphens without spaces. This is one of the three ways we put text inside 'brackets': using two commas, using two dashes and using a pair of round brackets. Two commas are the weakest and enclose asides or weak comments. Two hyphens – converted to *en dashes* by Word – suggest the enclosed words are saying something important. Round brackets enclose strong asides such as (see Figure 6).

You can also use a single *en dash* towards the end of a sentence – to emphasise a point. This is exactly the same as using two dashes as brackets described in the previous paragraph, except that a full stop acts as the closing bracket.

Preferably, aim to keep things simple and just use hyphens and en dashes. However, if you are using the American style, remember that it generally prefers the em-dash to the en-dash in normal text.

Another use for the *en dash* (in both British and American usage) is when specifying a range where it indicates 'to', or to express a connection where it indicates 'and'. Examples include: a London–New York flight (London to New York), and the author–reviewer relationship (author and reviewer). Notice that there are no spaces on either side of the en dash when used in this way.

American style guides, and some British ones, also require a closed *en dash* to express a range of numbers as in 2015–2020, 20–30 °C and pages 24–32. That might be fine for publishing houses, but most engineers would settle for a simple hyphen believing that their readers would neither notice nor care – but it is your choice. Most technical reports are sent from one department or organisation to another and publishing houses are not involved, and using hyphens is much easier than fiddling with closed-up en dashes, although substituting the word *to* may be even better as in 20 °C to 30 °C. The word *to* is especially helpful when expressing a range of negative numbers where hyphens, dashes and minus signs can be confusing. Use *to* as in, 'from -6 to -12' rather than 'from -6 – -12' or 'from -6 - -12'.

See also: Hyphen.

Data

Data is a plural word that is often misused as a singular. It takes a plural verb such as 'data are'. The true singular, *datum* looks worse than old-fashioned unless referring to a fixed reference point, such as a datum line or mark. You will rarely need to use the singular form but if you do, consider writing about 'one measurement' or 'one reading'.

Data, information

Both refer to knowledge. *Data* is the raw input, both the qualitative and quantitative attributes of variables such as measurements. *Information* is usable knowledge derived by processing data in some way, such as by structuring or analysing it.

Dates

Use the British or American style as appropriate: 11 January 2020 or January 11, 2020 (the latter with a comma). Using ordinal numbers is optional, such as 1st, 2nd and 3rd. Within the text, write the month in full, January not Jan. However, you can use abbreviations in tables and spreadsheets. Avoid the numbers-only style for an international readership because of the confusion between British and American practice. The date 12/10/20 means the 12 October in Britain but the 10 December in America. Include the year if there is a possibility that the report could be read in future years. There is an international standard (ISO 8601) which uses the format YYYY-MM-DD.

If you need to include the day of the week, use a comma after it as in 'Tuesday, 7 July 2020'. If this occurs mid-sentence, then use two commas as in 'On Tuesday, 7 July 2020, another incident occurred.'

Choose your style and use it consistently.

Decimate

Originally this meant (and still does) to reduce by 10 %, from its military origin to kill every tenth person. It is now also accepted as meaning to kill, remove or damage a large proportion of something. Avoid the word in reports because of this double meaning. Instead, prefer to state a proportion or percentage if appropriate.

Despatch, dispatch

See: Dispatch.

Different from, to, than

Prefer the traditional *different from* as it is used in both British and American English. *Different to* is common in British English but *different than* is common in American English.

Diffuse

Diffuse means to disperse, spread out, dilute or intermingle. For example, tiny quantities of other elements are diffused into pure silicon to make semiconductor devices. It does not mean render harmless; if you diffused a small quantity of arsenic into a pot of coffee it would not be harmless.

Disclaimer

Your organisation should stipulate whether a disclaimer is needed and how it should be phrased.

Discrete, discreet

Discrete means independent, separate or individually distinct parts, usually of the same type of thing. *Discreet* means being tactful, taking care to avoid embarrassing someone.

Dispatch, despatch

They mean the same thing but dispatch is now the more popular, especially in American usage. As always, be consistent.

Disinterested, uninterested

Strictly, *disinterested* means to be impartial or not influenced by something, whereas *uninterested* means a lack of interest. When conducting measurements or experiments you should be *disinterested*, in the sense of impartial, but not *uninterested*, in the sense of lacking interest. Because many people are not clear about the meaning of *disinterested*, consider using *impartial* instead.

Ditto

Do not use *ditto* in tables; complete each cell in full.

Document control/information

Your organisation should have rules about how to control documents. They will probably include issues about authority, such as the names of the author, reviewer and approver with dates; project information including the document number and issue date; a distribution list, including both internal and external recipients; a security classification if needed; and a document history stating version or revision numbers and dates.

Version numbering systems often use two numbers as in v1.2. The first number is the primary version number and the second signifies minor changes to it. Version 0.x refers to pre-publication or draft versions; v1.0 is the first published version; v1.1 indicates a minor change to v1.0; and v2.0 would be a major change to the original published version (v1.0). The software industry has set the standard to learn from and uses three numbers to represent major changes, minor changes and patches.

Double negatives

These must be avoided as they are confusing. A double negative is when two negative words in one sentence cancel out, just like when you multiply two negative numbers together. 'The evidence is not irrefutable.' 'The results are not inconclusive.'

Due to

See: Because.

Effect

See: Affect.

e.g.

The Latin *exempli gratia* means 'for example' and should not be confused with *i.e.*, which means 'that is'. Its correct form is *e.g.* with full stops. Some writers use it without full stops in informal writing. However, use the full stops in anything remotely formal such as a technical report.

See also: Latin words.

Ellipsis

An ellipsis (...) indicates that a word, words or even whole sentences have been left out. You may need an ellipsis in a report if you are quoting someone but leaving out part of what they wrote. 'A bird in the hand... two in the bush.' If the ellipsis comes at the end of a sentence there is no need for a full stop, but you will need a question mark if appropriate.

Emphasis

This means using some technique to emphasise certain words. For example, bullet points make lists stand out from the normal text – they *emphasise* the list. However, emphasis usually refers to using bold, italics, underline or capital letters. Many people overuse emphasis in their writing and the general advice for authors of technical reports is to make your phrasing provide the emphasis. That means you may have to think a little more about what you are trying to say.

If you do use emphasis, prefer italics in the text and bold for headings. Reserve underline for web addresses; elsewhere it is seen as old-fashioned as it was the only option available on most typewriters. Save all-capitals for possible use in major headings. Italics are sometimes used, as in this style guide, to highlight a word or a letter when discussing it. They are also used in references to highlight titles. Emphasis is discussed towards the end of Section 4.4 in Chapter 4 and references are discussed in Section 3.18 of Chapter 3.

Equations, formulas

Traditionally, equations and formulas are centre aligned; if you want to use left align then tab in by one tab. If appropriate, writing an equation within the text ($x = a/b$) can simplify the appearance and save space. The plural word *formulae* is still used in science, mathematics and academic writing although it seems to be giving way slowly to *formulas*. Choose one version and use it consistently.

et al.

Et alia is the Latin for 'and other people'. It is used in lists of references after the principal author's name to indicate that there are other, unnamed, authors. It is abbreviated to *et al.* with one full stop.

See also: Latin words.

etc.

Etc. is an abbreviation of the Latin, *et cetera* and it means 'and other things'. Use the full stop after it but do not add another one if the sentence ends with etc. Do not use *etc.* for 'and other people' where *et al.* is the correct term.

See also: Latin words.

Everybody, every body, everyone, every one

Everybody and *everyone* mean the same thing and mean 'every person' but they do not often appear in technical reports as they refer to people. *Every one* means 'each individual item' and is common in technical reports. It can also be used for 'each person'. You can think of *every one* as being 'every single one' or 'each of'. *Every body* refers to 'each body' as in a human or animal body, or where you are using 'body' to mean a group of people.

Everyday, every day

Prefer *every day* if you mean each day as in 'Measurements were taken every day for a month.' *Everyday* can also mean ordinary, typical or usual as in 'everyday clothes'.

Exclamation mark

Exclamation marks are used after exclamations which are rarely, if ever, used in technical reports. Will you use exclamation marks? No!

Experimental error

Experimental error is the difference between a measured (or estimated) quantity and its true value. There is always some inaccuracy in all measurements and the nearest to a true measure will be the mean or average of many measurements. How important experimental error is will depend on the subject matter of your report. Error analysis can be a complex discipline but often quoting the results plus or minus the estimated error is satisfactory, as in 240V ± 2V. Take care if writing percentages as 50 % ± 10 % is ambiguous; it could mean 45–55 % or 40–60 %.

Fact that

Very often, *the fact that* means nothing more than 'that' as in 'Given the fact that there were...', which means 'Given that there were...'

Fact, opinion

Distinguish clearly between facts and opinions. Your methods, results and analysis sections will be factual. Bring your professional opinion or judgement into the discussion section, especially when contrasting options. Your recommendations are also your professional opinions.

Farther, further

The traditional use of *farther* and *farthest* is in terms of distance such as in 'the farthest point'. However, *further* and *furthest* are now widely used in that sense. Use *further* and *furthest* when meaning extra or more advanced, as in needing 'further investigation'.

Fewer, less

These are often confused. *Fewer* means 'not as many'. Use it with countable nouns about things, objects, processes and so on that you can count, such as fewer than six, fewer computers and fewer measurements. *Less* means 'a smaller amount' or 'not as much' and is used with mass nouns, that is things that are uncountable (which does not mean they are unmeasurable). For example, 'less acid' or 'less crowded'. *Fewer* is for things you can count. *Less* is for things you cannot count.

Sometimes these guidelines are difficult to apply. The key question is whether you can count the items. Therefore, it is 'less milk' but 'fewer litres of milk', 'less cargo' but 'fewer containers'. 'If people drove fewer miles there would be less pollution.'

First, firstly

First and *firstly* mean the same thing as do *second* and *secondly*. In lists, the traditional *first...*, *secondly...*, *thirdly*, although odd, has been preferred although using *firstly... secondly..., thirdly* is also correct. The *-ly* ending looks old fashioned to some people, and it becomes cumbersome with higher numbers: seventhly, eighthly, ninthly. *Firstly* is also seen as a little more formal than *first*. The modern trend is towards simplicity where possible and many now prefer *first, second, third* and so on.

Flammable, inflammable, non-flammable

Flammable and *inflammable* mean the same thing with a slight preference for *inflammable* in British English and *flammable* in American English. Use the accepted term for your industry, bearing in mind that industrial practice may not agree with dictionaries. In some instances, public warning labels are an example, *flammable* may be preferred to *inflammable* to avoid uninformed people misinterpreting the *in-* as meaning *not*, as in *eligible* and *ineligible* which are opposites. The opposite to both terms is *non-flammable* (*nonflammable* in the USA).

Flesch reading ease

This measures the readability of a passage; the higher the score, the easier it is to understand. Try to achieve a score of around 45 to 60 across the entire report although you may struggle to get that high if you have a lot of jargon or long words. Try to score towards the higher end in the executive summary, discussion, conclusions and recommendations. As a guide, this paragraph has a score of 51.0, the Gettysburg address scores 65.0 and the children's rhyme *Mary had a little lamb* scores 87.4. See page 97 for a guide to what the scores mean.

See also: Section 5.1 in Chapter 5.

Flow of current

It is tempting to write about the *flow of a current* whether an electric current, river current or anything else that flows, but it is wrong to do so because a current is a flow. For example, 'a flow of electric current' means 'a flow of a flow of electric charge'.

Font

There are two styles of font: *serif* and *sans-serif* (or non-serif). Serifs are the small flourishes (fancy bits), on the ends of letters. Times New Roman is a popular serif font with flourishes while Arial is a popular sans-serif font without flourishes. Some organisations prefer a sans-serif font for headings and illustrations and a serif font for the normal text. Others see that as needlessly complicated. It seems to be accepted that serif fonts are more distinctive and therefore easier to read in good-quality print. Published books tend to use serif fonts.

Sans-serif fonts used to be recommended for reading on screen because of the lower resolution of screens compared to print, but screen resolutions have increased in recent years. So, it depends at least in part on the comparative resolutions of your printer and screen – or rather, your reader's printer and screen. Add into the mix that people are used to reading serif fonts in newspapers and books, and it gets even more complicated. At least, keep it simple by preferring one font and never use more than two.

Formulas

See: Equations.

Fractions

In Word, if autocorrect is switched on then typing, say, 1/2 (without spaces) will automatically give you the fraction ½. To get 2½, type 2 and leave a space before typing the fraction, then delete the space. For fractions that are not available you will have to settle for 13/16 and so on. Use hyphens if you write fractions as words, one-half, three-quarters, seven-eighths.

Frequent

See: Common

Full stop, full point, period

Full stops are also known as *periods* or *full points*. Use them to mark the end of a sentence including, in some styles, the final bullet point in a list written as a single sentence.

Omit them in initialisms, like the 'BBC', and it is now acceptable to omit them in common abbreviations, such as 'Mr', 'Mrs' and Dr. Other abbreviations end with a full stop. If the last character of a sentence is the full stop after an abbreviation, do not add a second full stop. (However, you should avoid using abbreviations in technical reports.)

Use full stops to mark the decimal point in numbers (34.67) although some countries use a comma (34,67). Also, use full stops when writing the time of day as in 14.30, although some American usage favours the colon, 14:30.

Geographical areas

These can be tricky. Capitalise the names of specific geographical areas, such as Northern Ireland, North America, South Africa and South East Asia (Southeast Asia in American usage). Do not capitalise when the name refers to a vaguely defined area, such as northern England or southern Scotland. Do not capitalise directions, such as head north or go south-east (but in the USA, 'southeast'). Always capitalise single letters that represent compass points: N, S, E, W.

Grammar checker

See: Section 5.1 in Chapter 5.

Heading styles

The most common way to show the hierarchy of headings in technical reports is to number them, normally using up to three levels. For example, a single number for the main section headings (sections 1, 2, 3), two numbers for subheadings (subsections 1.1, 1.2, 1.3) and three numbers for any sub-subheadings (sub-subsections 1.1.1, 1.1.2, 1.1.3). Three levels are enough for most reports although some organisations use four. If you really need four you could use a run-in heading for level four, where the first few words of the paragraph (in bold or italics) are used as the heading. If desperate for more, use letters in round brackets, (a), (b), (c), then roman numerals, (i), (ii), (iii), and possibly bullet points after that. But does anyone really need six or seven layers?

The numbering can be combined with indenting the second and third headings and their paragraphs. While this provides a clear hierarchy it also marches across the page, especially if there are four levels. This may not be too irritating on paper, but it can be a problem when reading online – especially on a phone. A run-in heading would not be indented further.

Emphasis is also used to show the levels of headings. The highest level could use all capital letters, although some style guides avoid that. The next level could use bold, and the third level could use italics. A run-in heading would use either bold or italics. Reserve underlining for links to websites.

Font size can also indicate hierarchy, the highest level having the largest font and the lowest level having the same size as the normal text. (Some styles use different fonts for headings and the body text, usually a sans serif font for headings and a serif font for the text.)

Even the writing style can be used to show the different heading levels. For example, ALL CAPITALS for Heading 1, 'Title Style' for Level 2 (Where all the Main Words Begin with a Capital Letter), and 'Sentence style' for level 3 (Where only the first word starts with a capital letter).

Spacing is also used to add emphasis to headings by adding more space above and below a heading than is used for the normal line spacing. The higher the heading level, the larger the spacing. Make the space above a heading greater than the space below it so that it clearly refers to what follows and not to what preceded it.

Left align all headings unless using the indented numbers approach, although some organisations centre the main section headings.

To help you to be consistent, use the built-in styles provided by the word processor but modify them to get the exact styles you want for each level.

See also: Section 2.9 in Chapter 2.

Hopefully

Avoid using the word *hopefully* for two reasons: first, you are writing a technical report that relies on factual evidence, logical analysis and professional judgement, not hope. Second, it annoys some people as there are some (nitpicky) arguments about what it really means.

It has two meanings, a traditional one (in a hopeful manner) and one that grew in the 20th century (it is to be hoped that). In a sentence starting, 'Hopefully, the recommendations will…' the first meaning is that the actual recommendations are hoping, which is silly and therefore wrong. The second meaning, which some reject as bad grammar, is that the author hopes or we hope that. Here, *hopefully* is called a sentence adverb and it modifies the entire sentence rather than a single verb. Regardless of this obscure argument, avoid the word.

However, nevertheless

Both introduce a contrast to whatever came immediately before. There is a slight difference with *nevertheless* being a little more formal and emphatic, possibly because it is used less often. You can start sentences with these words but you will usually need a comma after them.

Hyphen

Although hyphens and dashes look similar they are opposites in use; in general, with a few exceptions, hyphens join words together while dashes separate words.

The rules about using hyphens have been described as the most contradictory and volatile in grammar. Apparently (although I have not been able to find the original) another comment claims that if you take hyphens seriously you will surely go mad. Neither comment is very encouraging. In the hope this will not drive us mad, let us try to keep this relatively simple.

There are two types of hyphen: hard and soft. A *soft hyphen* is used to split a word at the end of a line. Avoid splitting words if you can. Splits are unnecessary when using left alignment. If the text is justified, then some imaginative rephrasing might avoid splitting words. The rest of this entry is about *hard hyphens*.

First, when two or more words are used together to give a single meaning (a compound term) there is always a question as to whether they should be separate, hyphenated or merged into a single word. Over time, two separate words tend to move from separate to hyphenated to joined up, but you cannot forecast how long that will take or even if they will ever join up. We used to write 'e-mail' but now we write 'email' and, apparently, we used to write 'down-stairs' although it must have lost its hyphen ages ago. There are many other examples of single words that were once hyphenated including postgraduate, lifelike, seaside and so on. There are also examples of words that were once hyphenated that are now two words including dining room and walking stick.

Unfortunately, because this depends on common usage there is no real rule and hence little consistency, so see what your dictionary recommends. Even then, dictionaries do not always reflect industrial practice and different dictionaries may give different advice. Perhaps this is what was meant in the above comment about hyphenation being contradictory and volatile.

With technical terms, follow the custom in your industry and keep a list of terms that you and your colleagues use and how you decide to write them. You may be lucky and find that other organisations in your sector have published their style guide on the internet, which Microsoft has done for instance. Even then, you may not agree with some of their decisions. Usually, do not hyphenate compound scientific terms, such as 'radioisotope', 'liquid crystal display' and 'potassium chloride solution'. As a guide, only hyphenate if the dictionary says so or if you believe it will help your readers and avoid ambiguity. Otherwise use a single word. Make your choice and apply it consistently.

Second, use a hyphen to merge two or more adjectives into a single compound adjective to describe a noun, such as in 'a small-scale study', 'an over-ride switch', 'six-year-old washers' and 'up-to-date records'. These form compound adjectives to describe the following noun. It is as if you are asking the reader to mentally join the words together to get the meaning.

Even using hyphens in this way is not consistent because if the expression in well known, or the meaning is abundantly clear, the hyphen can be omitted as in a 'first aid kit'. English is a living language so practice changes over time. Maybe one day we will have 'firstaid kits', but you cannot buy one yet and perhaps will never be able to.

Note that 'six year-old washers' has a very different meaning from 'six-year-old washers'.

It can become even more confusing when two hyphenated adjectives are moved from before the noun to after it, because one of them stops being an adjective and becomes a noun so the hyphen disappears. For example, 'a twentieth-century problem' could be written as 'a problem of the twentieth century'.

Third, words with prefixes, such as antistatic and subsection, generally do not use hyphens unless a hyphen helps to avoid confusion where there is a clash of letters as in de-ice, re-entry and ex-directory. However, in general, American practice prefers merging words to hyphenating them, at least more so than British practice, favouring deice and reentry but – ex-directory. Check in a dictionary, preferably online as you need it to be up to date (but up-to-date if preceding the noun, as in an up-to-date dictionary).

Fourth, all double-barrelled numbers from twenty-one to ninety-nine have hyphens when written out.

Fifth, use hyphens without spaces to show numbers or words that form a range, such as in 2015-2020, 20-30 °C, pages 24-31 and Monday-Friday. In American usage the longer *en-dash* without spaces is preferred and some British style guides agree: 2015–2020, 20–30 °C, pages 24–31 and Monday–Friday. Most technical reports are sent from one department or organisation to another and publishing companies are not involved; using hyphens is much easier than fiddling with closed-up en dashes, although substituting the word *to* may be even better as in 20 °C to 30 °C. The word *to* is especially helpful when expressing a range of negative numbers where hyphens, dashes and minus signs can be confusing. Use *to* as in, 'from -6 to -12' rather than 'from -6 – -12' or 'from -6 - -12'.

Sixth, use hyphens if you use the international system for writing dates: YYYY-MM-DD.

See also: Dashes.

Hypothesis, theory, law

Take care if you use these words in your report because they have specific meanings in science and technology but are used loosely elsewhere.

Think of a *hypothesis* as a possible explanation of something. It needs to be tested by making predictions based on it and testing those predictions by experiments, which will either support or disprove it. A *theory* is not a possible explanation, it is the accepted explanation of a phenomenon. It is the peak of scientific explanation that has been tested repeatably and quite likely refined along the way. It may have boundaries within which it is applied, and it is still being tested in the sense of always questioning whether new evidence supports it – because that is the scientific way. A *law* is different again as it is not an explanation; it is a statement of what is, not the how or why. *Laws* are usually relatively simple statements and are universal truths, often expressed mathematically.

i.e.

The Latin *id est* means 'that is' and should not be confused with *e.g.* which means 'for example'. Its correct form is *i.e.* with full stops. Some writers use it without full stops in informal writing. However, use the full stops in anything remotely formal such as a technical report.

See also: Latin words.

If... then

Put a comma after the *if* clause when that clause is an introduction to the main sentence. The comma, therefore, comes before the word *then*. The *if* clause sets a condition for the main part of the sentence. 'If you do that, then I will do this.' The comma is not always necessary but is usually preferred. For example, you may omit it if there is other punctuation nearby and you feel the sentence is becoming cluttered.

Illustrations

See: Section 4.10 in Chapter 4.

Impact

See: Affect.

Imply, infer

Writers *imply* something, which means they suggest something without explicitly stating it. Readers *infer* something from the message, which means they read something into it that is not explicitly stated. Writers of technical reports should be explicit in what they write. Your readers should not need to infer anything.

Inflammable

See: Flammable.

Information

See: Data.

Infrared, infra-red, ultraviolet

Use the single words, *infrared* and *ultraviolet*.

Initialisms

See: Acronyms.

In regard to

A minor point but some see this phrase as waffle. If it is the best expression then use it, but also try words like 'regarding', 'concerning' or 'about'.

In terms of

Another minor point that some see as waffle. Instead of writing, 'The results are questionable in terms of their accuracy,' try turning the phrase round to something like, 'The accuracy of the results makes them questionable.'

Internet, World Wide Web

Write *internet* with a lower-case letter *i*. Start each word of the *World Wide Web* with a capital letter, but *web* is all lower-case letters.

Inverted commas

See: Quotation marks.

Irritate

See: Aggravate.

Italics

Formally, *italics* have some specific uses in technical writing. These include variables in equations (but not the unit symbol) such as $t = 3$ s; foreign words including Latin but not i.e., e.g. or etc. (which are seen as English); for emphasis especially in lower-ranked headings; and in the titles of published books, journals and magazines (or you can use inverted commas/quotation marks).

Its, it's

These are often mistaken or written as typos. *Its*, without the apostrophe, means 'belongs to it' and is the possessive version of the pronoun *it*. It is the third part of the *his*, *hers*, *its* trio. *It's*, with the apostrophe, is the contraction of 'it is' or 'it has'. In technical reports you should not be using contractions and so should not need to write *it's* except perhaps in short email reports or in covering letters or emails.

Jargon

Jargon is the collection of special words, phrases, abbreviations, acronyms and initialisms used by specialists in a discipline. Some of that jargon will be known to your expected readers but some may not be. Avoid jargon as much as possible in the sections that non-specialists are likely to read: the executive summary, discussion, conclusions and recommendations. In the main body of the report your readers are most likely to be fellow experts and they should understand the jargon.

Justification

See: Alignment.

Latin words

You can use Latin words and expressions that have passed into the English language (e.g., i.e., via, etc., versus, sic, circa, which are written in normal/roman type – not italics) and any others that are common in your subject area. However, use the equivalent English words if you think some of your readers may not know them.

The following expressions are reasonably well known:

- *e.g.* (*exempli gratia*) means 'for example' but when using it you should provide at least two examples
- *i.e.* (*id est*) means 'that is', which refers to a strict equivalent almost like an equals sign
- *etc.* (*et cetera*) means 'and others' or 'and the rest'; do include the full stop
- *versus* means 'against' or 'opposed to', do not abbreviate it to *v* except in an illustration or table
- *sic* in round brackets inside a quotation indicates that the preceding word has an error that is in the original, such as a spelling mistake
- *circa* means 'about' or 'around' and is used with dates, usually historical
- *et.al.* (*et alia*) means 'and other people' and indicates that you have only named the first of two or more authors of a report, article or book
- *ibid.* (*ibidem*) means 'in the same place' and is used when citing or referencing a different part of the same publication as the last one cited or referenced
- *c.f.* (*confer*) means 'compare to' or 'see also'.

Law

See: Hypothesis.

Licence, license

Licence is a noun, usually a permit or certificate granting permission to do something. *License* is a verb that is used when granting permission. Both words are spelt *license* in American English.

Line numbering

Numbering the lines is not common in technical reports. However, it can be useful if several people will discuss and review the report together as it makes it easy to find specific phrases on a page. Also, it can be turned off before the report is submitted. In the Layout pane of Word, select the Line Numbers option, then select whether you want the numbers to be continuous or to restart at each new page or section. Then select the Line Numbering Options box, Layout, Line Numbers and define how you want them to appear such as every five or ten lines.

See also: Paragraph numbering.

Literally

Take care, do not use *literally* to emphasise or strengthen the next phrase. It means the next words should be understood in their literal or dictionary meaning. In a technical report, all your words should mean what they say and should not need that sort of emphasis.

Majority, most

Majority means more than half. *Most* can mean the same but usually means nearly all.

May

May refers to permission to do something; *can* refers to the ability to do something.

Me, myself

These are frequently confused, although you are unlikely to need either in a technical report. *Me* is the pronoun for when someone else is doing something to you such as in 'he treated me to an ice cream'. *Myself* is the pronoun to use when you do something to yourself, as in 'I treated myself to an ice cream'. The same logic applies to using *it* and *itself, you* and *yourself,* etc. While you will not use these pronouns in a technical report you may need to use them in a covering email or letter to a client. Please do not ask your client to 'send details to myself'. The 'self' pronouns are called 'reflexive' pronouns. Think of them as 'reflective' – they are used to reflect back onto the do-er: 'I' to 'myself', 'you' to 'yourself', etc.

Mean

See: Average.

Meaningful

Saying your results are *meaningful* is vague. Anyway, everything in your report should be meaningful. It is better to describe and explain what they mean. Having said that, *meaningful* can have a more specialised use in statistics, indicating that something is practically significant.

See also: Statistically significant.

Median

See: Average.

Mode

See: Average.

Most

See: Majority.

Must, should, could, would

The first three words are the three stages of a simple prioritisation method that some organisations use, in descending order: *must happen, should happen, could happen. Won't happen* is sometimes added as a fourth and lowest priority. The acronym MoSCoW is a memory jogger for the list. An alternative is: *must, should, could, don't.*

Both provide simple options for writing your recommendations and are alternatives to the high, medium, low or ABC systems of prioritising. A possible advantage is that the words convey clear action-orientated priorities. Sometimes in technical reports it may be better to express numerical probabilities if you can.

Would is the past tense of *will* and can be used to express a warning of future consequences of actions as in, 'Lowering the specification would risk having to backtrack in the future.' Using, 'will risk' is even stronger.

Myself

See: Me.

Nevertheless

See: However.

Nobody, no one

Nobody is one word, *no one* is two words without a hyphen. They mean the same thing and both take singular verbs.

Non-flammable

See: Flammable.

Number, numeral

Numbers are quantities or arithmetical values whereas *numerals* are symbols or words that we use to represent numbers. In a sense, numbers are ideas but we use numerals to write them.

Opinion

See: Fact.

Owing to

See: Because.

Oxford comma, serial comma

The *Oxford* or *serial* comma is the comma that appears before 'and' in the final item in a list of three or more items. It is used more in American English than in British English, although some British writers and publishers use it routinely. (Oxford University Press has used it for over a century.) It is a style issue and, while often unnecessary, it can sometimes help to avoid ambiguity. Decide whether to use it or not and then be as consistent as you can, but use it willingly if it will help the reader.

See also: Section 4.9 about lists in Chapter 4.

Paragraphs – average length

Long paragraphs are harder to understand than short ones. For standard A4 or US Letter pages, aim for an average paragraph length of, very roughly, four to eight lines over the entire report. Naturally, some will be longer and some will be shorter. Expect to have the longest average paragraph length in the heavy technical sections.

See also: Section 4.3 in Chapter 4.

Paragraph numbering

Some organisations like to number the paragraphs throughout the entire report or section by section. In Word, simply highlight the paragraphs you want to number and click the numbers button next to the bullet point button.

See also: Line numbering.

Parentheses

See: Brackets.

Per cent, percent, %, percentage points

Use the symbol when expressing a percentage with numerals (not words) such as in 25 %. Leave a space between the number and the symbol if following the SI style of writing numbers and units. When written as words, *per cent* is said to be more common in British English and *percent* more common in American English; choose and then apply consistently. *Percentage points* refers to the quantitative difference between two percentages so, while you can say that 25 % is 50 % (or half) of 50 %, it has 25 percentage points fewer than 50 %. Take care not to confuse the two and, as important, do not confuse any readers of the widely read parts of your reports who may not be as numerically literate as you are. If you need to write the number of a percentage in words, then write 'five per cent' or 'five percent'. Although in technical writing why not keep things simple and write '5 %' or '5%'?

Period

See: Full stop.

Personal pronouns

Pronouns should be avoided in all formal reports, apart from *it* and *they* when referring to things. For example, you can write 'when water boils, *it* will evaporate'. This guidance may not be as strict in informal reports, especially internal ones, but always check your organisation's style to see what is acceptable. All personal pronouns can be used in covering emails or letters.

Plurals

There are many ways to change a singular into a plural in the English language and most of them are wrong for any particular word. As you know, you cannot always add an *s*. You will know all the weird plurals for your own specialism but check in a dictionary for any specialist words you do not usually use.

Some words have a choice of plural, such as *formulas* or *formulae*, and *stadiums* or *stadia*. Where there is a choice, prefer the one you believe your readers are most familiar with. If in doubt, opt for the word that sounds more like English than Latin.

Practical, practicable

The meaning of *practical* is close to 'effective' or 'useful' and relates to real situations or actions whereas *practicable* is close to 'feasible' (which may be a better choice) and relates more to ideas or intentions.

Practice, practise

Practice is a noun whereas *practise* is a verb. You may need to *practise* taking these measurements because *practice* makes perfect. American English prefers *practice* for both.

Precision

Prefer precise statements to vague ones. If you mean, 'The obstacle was 10 m from the side of the road,' then write that and not, 'The obstacle was some way from the road.' Other vague words include few, many, rarely, frequently, occasionally, often, several, hot, cold and so on.

Principal, principle

Principal as an adjective refers to the 'main' or 'most important' component or ingredient of something. As a noun it refers to the person who leads certain organisations, such as the head or principal of a college.

Principle is a noun that refers to a fundamental truth or scientific law, such as Fermat's Principle or Heisenberg's Uncertainty Principle. It is also used for a code of conduct or moral principle.

Program, programme

The American spelling *program* is used in British English for software. For other purposes in British English use *programme* as in 'a programme of work'.

Proofreading

See: Section 5.3 in Chapter 5.

Punctuation

Treat punctuation as a form of customer service rather than a set of half-remembered rules. Punctuation is there to help your reader, not to satisfy a devotee of English grammar. That said, some uses of punctuation are right and some are wrong, and this guide includes comments about all the punctuation marks. Remember that most short sentences need little punctuation, which means they are easier to write as well as easier to read. When faced with a difficult punctuation puzzle, it may be quicker and easier to rephrase the sentence than solve the puzzle. Some dictionaries have a guide to punctuation as an appendix.

Question mark

Sometimes in technical reports, the writer states the aim as a direct question which then ends with a question mark instead of a full stop. Do not use a question mark if the question is implied rather than stated.

Quotation marks

Occasionally you may want to quote from other work. The fashion in British English is said to be to use *single quotation marks* for the main quotation and *double quotation marks* for a quotation within a quotation. The reverse is usual in American English. In British English this is not a rule and many people do the opposite, which is common in newspapers and magazines. It matters little so, as usual, decide and then be consistent.

Using other punctuation with quotation marks can be tricky. When a quotation is part of a larger sentence, place any punctuation marks that belong to the quotation inside the closing quotation mark. Place those that belong to the larger sentence outside the quotation marks – think of how you use brackets in mathematics. In American usage, put the punctuation marks inside the closing quotation mark regardless of whether they belong to the quotation or to the larger sentence. An example will make this clearer.

British style: Note that 'six year-old washers' has a very different meaning from 'six-year-old washers'.

American style: Note that 'six year-old washers' has a very different meaning from 'six-year-old washers.'

However, when the final part of a larger sentence is a quotation that itself is a complete sentence, the quotation's full stop also serves to end the larger sentence. A second full stop is not needed after the final quotation mark. This is logical but in British usage it can produce what, at first sight, appear to be anomalies because when the quotation is not a complete sentence the full stop comes outside the final quotation mark. An example makes this clearer.

This is correct in both British and American usage:
As the report said, 'There are many anomalies.'

But:
British style: 'There are,' the report said, 'many anomalies'.
American style: 'There are,' the report said, 'many anomalies.'

On a separate note, while quotation marks can be used to highlight a new or unusual word do not use them because you cannot think of the right word. (At this stage of the investigation I also 'touched based' with other departments.) Find the correct word and use it.

Finally, quotation marks are also known as *inverted commas*.

The most important consideration is to choose a style and use it as consistently as you can.

Readability statistics

See Section 5.1 in Chapter 5.

Reason is because

Many people do not like the phrase *the reason is because* as it can be a duplication where *the reason is* means *because* and you get 'because because'. This may be pedantic but avoid the risk of irritating any of your readers who see it as bad grammar and try, 'the reason is that' instead.

References

See: Section 3.18 in Chapter 3.

Regular

See: Common

Say, state

State is slightly more formal that *say*. Use *state* in the sense of expressing fully, clearly or formally.

Semicolon

Semicolons can separate lengthy items in a list and are essential if the lengthy items themselves need commas. They are no longer necessary at the end of bullet points, but you can use them if you wish. Semicolons can also merge two short sentences into one when they complement or parallel each other, instead of using a conjunction such as *and*. Using a conjunction may be simpler.

Sentences – average length

Generally speaking, short sentences are easier to understand than long ones. Use the readability statistics to check the average sentence length section by section, aiming for around 17 to 22 words per sentence over the whole report. Aim for a little less in the widely read sections and a little more in the heavy technical sections.

See also: Section 4.4 in Chapter 4 and Section 5.1 in Chapter 5.

Serial comma

See: Oxford comma.

Shall, will

This is an old conundrum. It is said that traditionally in England, *shall* has been used for the first-person pronouns (*I* and *we*) and *will* has been used for the second and third person pronouns (*you, he, she, it* and *they*) whereas, elsewhere, particularly in Scotland, Ireland and America, it was the reverse, *will* for the first person and *shall* for the second and third. Reversing any of them added more emphasis when giving an instruction. Some authorities argue that it was never that clear cut. This grammatical argument rumbles on although many people now prefer to use *will* with all the pronouns, reserving *shall* for when they want extra emphasis.

See also: Personal pronouns.

Slash, solidus, stroke

Avoid using this in the text although it may be useful in tables and illustrations. In text, if you mean *or* then write *or*. If you do use the *solidus*, then it should not have a space before or after it, as in web addresses.

See also: And/or.

Spaces at the end of sentences

Nearly all style guides now stipulate one space at the end of a sentence. Two spaces were common in the age of the typewriter to make the spacing clear but are no longer needed.

Stationary, stationery

Stationary means a lack of movement. *Stationery* refers to paper and envelopes. As a memory jogger, remember 'e for envelopes'.

Statistically significant

The term *statistically significant* has passed from its use in statistics to a more widespread use. It is often misused to stress 'importance' or 'significance'. Its correct meaning is that something is unlikely to occur by chance. Whether that something is important or not is another matter. Take care if you use the phrase.

See also: Meaningful.

Style

Style refers to the way the author plans, writes and edits the report. It is about your choice of words, the length of the sentences you use, the balance you strike between active and passive sentences, the amount of nominalisation you use, how you structure your paragraphs and sentences and a host of other things.

See also: Tone.

Superfluous words

When speaking, we often use common expressions that we have heard many times and they frequently contain too many words. These are easily carried over into our writing and need to be weeded out when editing. Eliminate expressions like 'plan in advance' (plan), 'the main essentials' (essentials), 'few in number' (few), 'brief in duration' (brief), 'active consideration' (consideration) and so on. My favourite bad example from a technical report is: 'irreducible minimum'.

Symmetry

Symmetry adds a nice touch to writing and can help sentences to flow. For example: '*Giving* a presentation is harder than *to write* a report', does not flow well because one verb is a present participle while the other is an infinitive. Making them both the same style improves the flow: '*Giving* a presentation is harder than *writing* a report,' or 'It is harder *to give* a presentation than *to write* a report.'

Systematic, systemic

You are far more likely to use *systematic* than *systemic* in a technical report. If something is *systematic* then it is organised in some way, there is a system to it. *Systemic* refers to something that affects the whole system or body of something. The term is used in biology and medicine. Also, reporters use it to criticise a *systemic failure* of an organisation, that is a failure that runs through the entire organisation.

Tables

See: Section 4.11 in Chapter 4.

That, which

The Microsoft Grammar Checker will often flag a possible error where *that* and *which* may seem to be interchangeable. The distinction lies in recognising two distinct types of expression, one that defines and one that describes.

Defining clauses, which define the subject, should use *that* although in British English *which* is often used. For example, 'The spares *that* arrived on time were used.' This implies that other spares did not arrive on time and were not used. The words 'arrived on time' define which spares we are talking about and are crucial to the meaning of the whole sentence.

Describing clauses merely add extra information that is not essential to the basic meaning of the sentence. For example, 'The spares, *which* arrived on time, were used.' This means that all the spares were used and, by the way, they all arrived on time. The words 'arrived on time' are now merely descriptive and without them the basic meaning of the sentence (the spares were used) would still be correct. Here, the correct word is *which*. Critically though, two commas enclose the descriptive words as if they are grammatical brackets. The commas must be there.

Here is an easy way to remember the rule. If the clause is essential to the meaning, use *that* without commas; if it is not essential, then use *which* with commas. For the reader, the presence or absence of the commas is the most important clue.

Several style guides give this clever example, although I do not know who thought of it first. 'This is the house *that* Jack built; but the other house, *which* John built, is falling down.'

Theory

See: Hypothesis.

Time of day

Prefer to use the 24-hour clock when writing times as it is unambiguous. You can use either a full stop or a colon to separate the hours and minutes, e.g. '16.30' or '16:30'. British usage favours the full stop whereas American usage favours the colon. The international system (ISO 8601) requires colons – hh:mm or hh:mm:ss, so 16:30 or 16:30:15 if seconds are needed. (Note, ISO 8601 requires hyphens in dates: YYYY-MM-DD.)

If you decide to use a.m. and p.m. then prefer to write them in lower-case letters with full stops and leave a space after the number, for example 3.00 p.m. Some writers omit the full stops, some omit the space and some omit both. Make your decision and, as usual, apply it consistently. Try to match it to your choice of style for e.g., eg, i.e., ie, 5 kg, 5kg, etc. Some American usage prefers capital letters, as in 3.00 P.M. or 3:00 P.M.

Tone

Tone is about how your writing style affects your readers, how they feel about it and the impression it makes on them. Do they see it as authoritative and trustworthy? You can think of style as the cause and tone as the effect.

See also: Style.

Ultraviolet

Spell ultraviolet as a single word.

Uninterested

See: Disinterested.

Unique

Unique does not mean rare, it means the only one of its kind. Prepare to be criticised if you use *unique* to emphasise rarity as in 'the conditions were unique', when you meant that the conditions were 'rare' or even merely 'unusual'.

Upper-case letters

See: Capital letters.

Verb tenses

Verbs have three main tenses *past*, *present* and *future*, and each main tense has four aspects: *simple*, *continuous*, *perfect* and *perfect continuous*. This gives twelve tenses. Knowing these names is not important to most engineers but using them correctly is important to your readers. Here is an example of each:

Past simple: It oscillated
Past continuous: It was oscillating
Past perfect: It had oscillated
Past perfect continuous: It had been oscillating.

Present simple: It oscillates
Present continuous: It is oscillating
Present perfect: It has oscillated
Present perfect continuous: It has been oscillating.

Future simple: It will oscillate
Future continuous: It will be oscillating
Future perfect: It will have oscillated
Future perfect continuous: It will have been oscillating.

The tenses you are most likely to use in each section of a technical report are mentioned in Chapter 3.

Which

See: That.

While, whilst

They mean the same thing, although *whilst* is not used much in American English. To some readers, *whilst* can seem a little more formal than *while*.

Will

See: Shall.

World Wide Web

See: Internet.

X-rays

This is normally hyphenated and uses a capital X, *X-rays*

BIBLIOGRAPHY

This bibliography lists the main style guides I have consulted while writing this book. I am indebted to them and to various organisations, including some of my clients, who have allowed me to consult their in-house style guides. Also, the list includes some older books that have guided me over the years. The web pages were checked and accessed in February 2022.

Bureau International des Poids et Mesures. 2019. *SI Brochure: The International System of Units.* 9th ed.
https://www.bipm.org/en/publications/si-brochure/

The Chicago Manual of Style Online. 2017. 17th ed. Chicago: University of Chicago Press.
https://www.chicagomanualofstyle.org/home.html

Council of Science Editors. 2014. 8th ed. *Scientific Style and Format Online.* Chicago: University of Chicago Press.
https://www.scientifcstyleandformat.org/home.html.

Cutts, M. 2020. *Oxford Guide to Plain English.* 5th ed. Oxford: Oxford University Press.

DISC. 2001. *Business Documents – Guidance on effective layout and presentation, PD 0017:2001.* London: British Standards Institution.

The Economist. 2018. *The Economist Style Guide.* 12th ed. London: Profile Books Ltd.

Gowers, E. 2015. *Plain Words.* Revised and updated by Rebecca Gowers. London: Penguin.

The Guardian. 2022. *Guardian and Observer Style Guide*
https://www.theguardian.com/guardian-observer-style-guide-a

Strunk, W and White, EB. 2000. *The Elements of Style*. 4th ed. New York: Longman.

Truss, L. 2003, 2007. *Eats, Shoots and Leaves*. London, Profile Books.

UK Government Digital Service. Updated 2020. *Style Guide*.
https://www.gov.uk/guidance/style-guide

University of Oxford. 2016. *University of Oxford Style Guide*. Oxford: Oxford University Press.
https://www.ox.ac.uk/sites/files/oxford/media_wysiwyg/University%20o
f%20Oxford%20Style%20Guide.pdf

Waddingham, A, ed. 2014. *New Hart's Rules*. 2nd ed. Oxford: Oxford University Press.

Zinsser, W. 2006. *On Writing Well, 30th Anniversary Edition*. Collins.

ABOUT THE AUTHOR

Tony Atherton is a trainer and writer who sees himself as semi-retired. After graduating with a degree in electronics and a PhD, he worked in the electrical and electronics industry for nearly 30 years.

This included time in the Royal Navy, with GEC-Marconi, as a lecturer at the University of Hong Kong, with the UK's Independent Broadcasting Authority, and as the training manager at NTL – now part of Virgin Media. He became a freelance trainer and writer in 1997.

A former Chartered Engineer (CEng) he still runs courses for engineers on writing technical reports. His clients have included the UK's Health and Safety Executive (HSE), Panasonic, Thales, Raytheon and lots of others including large engineering consultancies.

He has written nearly a hundred articles for various technical magazines and has had four books published by traditional publishers.

OTHER BOOKS BY THIS AUTHOR

Report Writing for Professionals

A sister book to this one. It is aimed at professional people working in disciplines other than science, engineering and technology. It omits the specifically technological aspects covered in this book.

30 Minutes Career Fast Track Kit - Market Yourself

This unique pack contains four best-selling titles from Kogan Page's popular 30 Minutes series that together represent the complete guide to achieving career success. Tony wrote one of the four books: 30 Minutes to Market Yourself.

30 Minutes to Manage your Time Better

There are 168 hours in every week but some weeks it feels as if we have been short-changed. Where did the time go? This text, 30 Minutes to Manage Your Time Better, advocates spending 30 minutes each day taking control of your time. Published by Kogan Page.

How to be Better at Delegation and Coaching

Delegation and coaching are skills that complement each other. This text provides hints and tips on developing these skills to improve performance in the organisation. It covers the processes involved in good delegation and coaching and provides an action list of how to improve both skills. Published by Kogan Page.

From Compass to Computer: A History of Electrical and Electronics Engineering

Although the history is well known to historians of science and technology it is relatively unknown to the majority of practising engineers. This book aims to bring them a readable account of how their subject developed from its early days as the two primitive sciences of electricity and magnetism to the 20th Century's vast engineering applications. Published by Macmillan Press and San Francisco Press.

www.ingramcontent.com/pod-product-compliance
Lightning Source LLC
Chambersburg PA
CBHW030637220526
45463CB00004B/1558